Steviol Glycosides

Cultivation, Processing, Analysis and Applications in Food

Food Chemistry, Function and Analysis

Series editors:
Gary Williamson, *University of Leeds, UK*
Alejandro G. Marangoni, *University of Guelph, Canada*
Juliet A. Gerrard, *University of Auckland, New Zealand*

Titles in the series:
1: Food Biosensors
2: Sensing Techniques for Food Safety and Quality Control
3: Edible Oil Structuring: Concepts, Methods and Applications
4: Food Irradiation Technologies: Concepts, Applications and Outcomes
5: Non-extractable Polyphenols and Carotenoids: Importance in Human Nutrition and Health
6: Cereal Grain-based Functional Foods: Carbohydrate and Phytochemical Components
7: Steviol Glycosides: Cultivation, Processing, Analysis and Applications in Food

How to obtain future titles on publication:
A standing order plan is available for this series. A standing order will bring delivery of each new volume immediately on publication.

For further information please contact:
Book Sales Department, Royal Society of Chemistry, Thomas Graham House, Science Park, Milton Road, Cambridge, CB4 0WF, UK
Telephone: +44 (0)1223 420066, Fax: +44 (0)1223 420247,
Email: booksales@rsc.org
Visit our website at www.rsc.org/books

Steviol Glycosides

Cultivation, Processing, Analysis and Applications in Food

Edited by

Ursula Wölwer-Rieck
Friedrich-Wilhelms-Universität Bonn, Germany
Email: woelwer-rieck@uni-bonn.de

ROYAL SOCIETY OF CHEMISTRY

Food Chemistry, Function and Analysis No. 7

Print ISBN: 978-1-78262-830-9
PDF ISBN: 978-1-78801-055-9
EPUB ISBN: 978-1-78801-566-0
Print ISSN: 2398-0656
Electronic ISSN: 2398-0664

A catalogue record for this book is available from the British Library

The Royal Society of Chemistry is a charity, registered in England and Wales, Number 207890, and a company incorporated in England by Royal Charter (Registered No. RC000524), registered office: Burlington House, Piccadilly, London W1J 0BA, UK, Telephone: +44 (0) 20 7437 8656.

Visit our website at www.rsc.org/books

Printed in the United Kingdom by CPI Group (UK) Ltd, Croydon, CR0 4YY, UK

Preface

In recent years, high-intensity sweeteners have become more and more important as they are intended to face the worldwide problem with excessive sugar consumption which is known to be directly related to metabolic disorders, obesity, diabetes and other diseases. Therefore, intensive sweeteners are nowadays widely spread in foodstuffs with growing demand in beverages. But the use of sweeteners as food additives is controversially discussed and there are concerns mainly about the safety of synthetic sweeteners. For example, the safety of aspartame, a widely used sweetener especially in beverages, is permanently discussed despite of global approvals and recent reevaluations, *e.g.* by the European Food Safety Authority (EFSA) in 2017. Therefore, a sweetener like steviol glycosides colloquially called "stevia", which is of natural origin and very stable during food production is of rising interest. It derives from the plant *Stevia rebaudiana* Bertoni and is applied in food, feeds and cosmetics as a white powder with a purity of more than 95%.

Due to their natural origin steviol glycosides have a lot of interesting aspects being worthy to be elucidated. Steviol glycosides represent a class of diterpenes with more than 40 different single components. The steviol glycosides rebaudioside A and stevioside are most abundant in the plant. Obviously, the composition of the different sweet compounds in the plant can vary significantly depending on several influences like genotype, climate, soil composition *etc.* The taste of the sweet mixture is also varying decisively depending on its steviol glycoside composition. It is obvious that a lot of aspects have to be taken into consideration for gaining a well tasting and safe high-purity sweetener.

In this book, we try to reflect these aspects by discussing acceleration and improvement of the traditional breeding of the stevia plant, traditional and new strategies to enrich, isolate and purify the interesting sweet ingredients

Food Chemistry, Function and Analysis No. 7
Steviol Glycosides: Cultivation, Processing, Analysis and Applications in Food
Edited by Ursula Wölwer-Rieck

Published by the Royal Society of Chemistry, www.rsc.org

in commercial scaled-up processes, novel trends in the analysis of steviol glycosides, further constituents in the plant and strategies for their analysis and the potential antioxidant capacity of the stevia extracts. A precondition for improvements in breeding is profound knowledge of the biosynthetic pathways for the synthesis of the sweeteners in the plant. Furthermore, it is essential to reflect upon potential adverse health effects which have to be considered together with metabolic functions of the steviol glycosides in the human body. An example for non-food application of steviol glycosides is their use in dentistry, which is discussed. Last but not least, we put a special emphasis on the most important aspect deciding the economic success of the sweetener: we look at its taste. After an introduction about taste perception in general, hints were given that and how steviol glycosides can influence both taste perception and insulin secretion.

I would like to thank the esteemed chapter authors for being part of this book and the publisher Royal Society of Chemistry, namely Nicki Dennis for the idea and Katie Morrey for speeding up the completion of the project. Special thanks are extended to Benno Zimmermann who offered numerous helpful suggestions and corrections.

We hope that this book will be a gain for students, teachers and researchers with interests in stevia as a sweetener of natural origin.

Ursula Wölwer-Rieck

Contents

Food Chemistry, Function and Analysis No. 7
Steviol Glycosides: Cultivation, Processing, Analysis and Applications in Food
Edited by Ursula Wölwer-Rieck

Published by the Royal Society of Chemistry, www.rsc.org

CHAPTER 1

Crop and Steviol Glycoside Improvement in Stevia by Breeding

S. TAVARINI, B. PASSERA AND L. G. ANGELINI*

Department of Agriculture, Food and Environment (DAFE), University of Pisa, Pisa, Italy
*E-mail: luciana.angelini@unipi.it

1.1 Introduction

Worldwide demand for both stevia leaves and purified steviol glycosides (SVglys) is steadily increasing,[1] and it is expected to increase further in the future, as metabolic disorders such as type-II diabetes and obesity are becoming ever more prevalent. Despite the stevia global market size being characterised by rapid progress, the agricultural production of this crop is still problematic and insufficient to meet the growing global demand. Stevia yields still remain low and unstable in many countries, where the crop is of recent domestication and/or performed by small- and medium-sized farmers, because of high input costs (in particular for planting and establishment), lack of suitable adapted and available varieties, limited expertise in the cultivation, poor disease control and lack of irrigation. To improve the

Food Chemistry, Function and Analysis No. 7
Steviol Glycosides: Cultivation, Processing, Analysis and Applications in Food
Edited by Ursula Wölwer-Rieck

Published by the Royal Society of Chemistry, www.rsc.org

competitiveness of stevia production it is important to produce improved, higher-performing and yielding crops, that are more resistant to drought, extreme weather conditions and a range of biotic stresses.

Therefore, genetic improvement with the development of varieties with higher leaf and SVgly yields and greater resistance to abiotic and biotic stresses, in comparison with the most and currently known cultivars, are important goals in stevia breeding.

1.2 Objectives of Stevia Selection and Breeding

The general objectives of stevia selection and breeding are: (1) to increase the leaf yield and its stability, (2) to ensure high quality value in terms of SVgly content and composition, and (3) to produce types with greater resistance and resilience against disturbance and stress, and targeted to specific growing conditions and farming needs.

The main desired traits, to reach through breeding program strategies in stevia, can be summarised as:

- high leaf yield per unit area
- high leaf-to-stem ratio
- rapid growth rate and regrowth capacity
- enhanced photosynthetic activity
- high SVgly content in the leaves
- high content of specific SVglys (*e.g.* rebaudioside A – Reb A)
- high adaptability to a wide range of pedoclimatic conditions
- resistance to environmental stresses, pest and diseases
- photoperiod insensitivity
- self-compatibility for viable seed production.

1.2.1 Dry Leaf Yield per Plant and Unit Area

Different characteristics are associated with dry leaf yield such as leaf thickness, number of leaves per plant, leaf surface area, number of branches.[2–5] *S. rebaudiana* leaves are relatively small, simple, opposite, subsessile, with large variability both in shape and size, ranging from oblanceolate to lanceolate or ovate.[6] It is possible to select high yielding varieties or hybrids characterised by longer and broader leaves, higher specific leaf area and thickness. The plants should have compact stature and green erect stems, upright and multiple branched. A reduced apical dominance with a higher number of branches, instead of a long slender plant with a single shoot, is preferred. Stem internodes should be closely spaced, to accommodate more leaves on the stems. Plants bearing upright leaves, instead of horizontal, have a better interception and utilisation of light, and are preferred in high plant density crop cultivation.

1.2.2 Leaf-to-stem Ratio

The stems of stevia plants do not contain appreciable quantities of SVglys, which are mainly accumulated in the leaves.[7,8] Hence, varieties that have a higher leaf-to-stem ratio are desirable because they yield higher quantities of SVglys per unit of harvested plant biomass.

1.2.3 Growth and Photosynthetic Activity

Equally relevant, is a higher relative growth rate and a better capacity to regrow after each harvest, giving new leafy shoots rapidly, thus allowing multiple harvests per year. Other desirable agronomic characteristics include lodging resistance.

Dark green leaves, with a higher chlorophyll content and photosynthetic rate per unit leaf area, that show better efficiency in converting solar energy in leaf biomass and SVglys is another important characteristic[9] that should be considered in breeding strategies.

1.2.4 Wider Crop Adaptability, Yield Stability and Abiotic Stress Resistance

Although originating from the Paraguayan highland of Amambay, between 22° and 25° S latitude and 55° to 56° W longitude,[10–15] stevia can be grown over a wide range of climatic conditions, from semi-humid, subtropical to temperate zones.[16] Once established, it can tolerate mild frosts, but not long periods of hard frosts, which kill the roots of the plant. This lack of winter hardiness means that stevia can be grown as a perennial (three to five years) in temperate to warm climates but as an annual in colder regions.[16]

Despite its suitability for general cultivation over a wide range of environmental conditions, a wider adaptability with particular attention to cold and drought resistance, is still considered to be an important objective in stevia plant breeding, because it helps in stabilising the crop production over regions and seasons.

Stevia plants suffer from several abiotic factors such as drought, extreme temperatures like heat, cold and frost, wind, flooding, soil salinity, excessive light intensity and UV exposition. Stevia plants vary considerably in their tolerance to abiotic stresses. The development of drought-resistant cultivars of stevia through selection and breeding is of considerable economic value for increasing crop production in areas with low precipitation or without proper irrigation. The development of drought tolerance in plants is the result of the overall expression of many traits in a specific environment, including both drought escape (rapid development to complete a life cycle before drought) and drought avoidance (reducing water loss to prevent dehydration) strategies. Water stress reduces photosynthesis in the leaves of stevia due to stomata closure in the short-term, and due to photo-inhibition damage and to the inactivation of RuBisCO enzyme (ribulose-1,5-bisphosphate carboxylase/oxygenase) in the long-term.[9] In this context, physiological and biochemical traits that contribute to enhance water use efficiency and photosynthesis at an early stage can be very useful. In semi-arid regions, resistance/tolerance to salinity is one of the major priorities to be addressed. Salinity reduces plant growth and yield, as well as SVgly content of the leaves. Since stevia shows variability in what concerns the sensitivity or tolerance to salt stress,[17–19] future research should focus on molecular, physiological and metabolic aspects of this stress tolerance to facilitate the development of cultivars with an inherent capacity to withstand salinity stress.

In temperate and cool regions, low temperature is the primary abiotic stress that limits crop productivity, since stevia is very sensitive to hard frosts especially in the first year of growth. The possibility to select superior over-wintering varieties makes it possible to grow stevia as a perennial, also under mid- to high-latitude regions. This contributes to both enhance the productivity and to reduce cultivation costs for planting. Recent breeding programs have targeted the development of cultivars that can withstand multiple stresses by assembling series of genes from different parents into a single genotype by a phenomenon called gene pyramiding.

1.2.5 Biotic Stress Resistance

Crop plants are attacked by various diseases and insects, resulting in considerable yield losses. Genetic resistance is the cheapest and the best method for minimising such losses. Fortunately, there are few pests and diseases recorded in stevia, except for a limited number of fungal diseases. The fungal diseases Septoria leaf spot (*Septoria steviae*), Alternaria leaf spot (*Alternaria alternata*), stem rot (*Sclerotium delphinii* Welch.), root rot (*Sclerotium rolfsii*), powdery mildew (*Erysiphe cichoracearum* DC), damping-off (*Rhizoctonia solani* Kuehn.) and *Sclerotinia sclerotiorum* have been reported.[5,20–25] *Septoria steviae* and *Sclerotinia sclerotiorum* are the two most damaging and widespread fungi,[1,26] to be reported in stevia grown in Canada,[26] India,[27] and Italy.[28] There is a need to develop resistant or tolerant varieties through the use of resistant donor parents available in the gene pool.

1.2.6 Photoperiod Insensitivity

The biological cycle of *S. rebaudiana* is photoperiod-dependent and latitude is the major factor affecting plant reproduction. Flowering in stevia has adverse effects on SVgly yield. In fact, the optimal time to harvest the leaves is at the onset of flowering, when the accumulation of SVglys reaches its peak.[16,29–31] When a stevia plant flowers, its SVgly content decreases. Stevia is a short-day plant with a critical day length for flowering of about 13 h.[32] Thus, the varieties with less sensitivity towards variation in day length are less prone to flowering, even when grown at medium–high latitudes. This is a very important desirable characteristic in stevia, therefore, selection may identify photo-insensitive varieties in the future.

1.2.7 Self-compatibility for Viable Seed Production

Stevia is characterised by a complex mechanism of seed multiplication. The species is hermaphroditic, highly cross-pollinated, photoperiodically sensitive, and produces tiny flowers in small capitula with five white tubular flowers.[5,33] These flowers have sporophytic self-incompatibility.[33–35] Seed yield and poor germination ability is one of the major constraints caused by self-incompatibility.

S. rebaudiana produces two types of achenes, black and tan coloured.[35–38] The black type is produced by cross-pollination and is characterised by high-germination, while the tan achenes, which do not have the seed embryo, are originated from self-pollination.[35] Even if self-incompatibility is a genetic mechanism, which prevent and thus encourage allogamy, it can be problematic for breeding techniques to be employed for crop improvement. The possibility to select self-compatible lines enables more efficient breeding techniques, and to select cultivars to be easily reproduced by seed.

1.2.8 Steviol Glycoside Composition and Yield

More than 100 compounds have been identified in *S. rebaudiana*, the best known of which are the SVglys, particularly stevioside (Stev) and Reb A, being the most abundant.[39] SVglys are tetracyclic ent-kaurene diterpenoids and share four steps of their biosynthesis with the gibberellin pathway.[40–42] The accumulation of SVglys occurs in active photosynthetic tissues (mainly in leaves). Therefore, quality is related to the improvement of SVgly content and of specific SVglys (namely Reb A, D and M) as a measure of taste. A higher content of antioxidant compounds is important, too.

1.2.8.1 Higher Steviol Glycosides Content in the Leaves

The SVgly content in *S. rebaudiana* leaves varies according to genotype, phenological stage, and growth conditions.[5,26,43,44] The original wild varieties had a 2–3% total SVgly content of leaf dry mass.[5] Breeding and selection have produced varieties with higher total Stev content up to 16–20% of leaf dry matter compared with the common lines of 8–10%.[45] The total SVgly content may be further improved up to 30%.[5]

1.2.8.2 Higher Content of Specific Glycosides

Stevia accumulates more than 30 SVglys in varying concentrations. The best known include Stev, Reb A–G (Reb A–G), steviolmonoside, steviolbioside, rubusoside, dulcoside A (Dulc A).[46–48] The two major SVglys, Stev and Reb A, account for more than 90% of the total SVglys found in stevia leaves.[39,48,49] Stev is estimated to be 110–270 times sweeter than sucrose but has a well-known liquorice off-taste and lingering sweet aftertaste, which reduces its acceptability. Reb A is reported as being 140 to 400 times sweeter than sucrose and more pleasant-tasting than Stev.[50,51] The quality, as a measure of taste, is usually defined as the ratio of Reb A to Stev. If Reb A is present in equal quantities to Stev, it appears that the aftertaste is eliminated.[5] Therefore, plants with a higher proportion of Reb A and less Stev are required to improve taste. Genotypes/lines descriptions indicate that sufficient genetic variability exists to make significant genetic gains in Reb A content and the RebA-to-Stev ratio.[43,52–55]

Morita and Yucheng[56] developed a variety of *S. rebaudiana*, by repetitive breeding and selection, which contained 2.56 times more Reb A than Stev. Nowadays, there are several varieties that can have Reb A content up to 80% of the total SVglys, with very little bitterness and aftertaste.[4,56] Other SVglys, besides Reb A, are of interest for their good taste. Among these, of particular interest are Reb D and Reb M (also known as Reb X), which have a more desirable taste profile than Reb A, even when present in tiny quantities (less than 0.5% by dry leaf weight).[57,58] It has been reported that Reb D, at the same concentration, has less astringency and bitterness taste perception attributes than Reb A, and behaves more like sucrose in terms of sweetness properties.[59] Therefore, there is great interest in increasing the Reb D concentration in stevia leaves, developing improved varieties. Stevia varieties containing high levels of Reb D have been recently developed, for example, the patented variety named "Stevia plants with an increased rebaudioside D content" (patent no. WO 2014146084 A1, Table 1.1).

In January 2016, Reb M was positively approved by EFSA's Panel on Food Additives and Nutrient Sources and has been included among the 10 SVglys approved for use in Europe. After this approval, high-purity Reb M is marketed for food and beverage applications.[58]

1.2.8.3 Higher Content of Specific Not-steviol Glycosides Compounds

Beyond SVglys, stevia leaves contain other important constituents, such as minerals, vitamins, phenolic compounds (mainly hydroxycinnamic acids and flavonoids), alkaloids, water soluble chlorophylls and xanthophylls, with potential beneficial effects on human health.[48,60–64] Flavonoids (flavonols and flavones) are particularly important for their high antioxidant capacity,[48,65] as well as phenolic acids, like chlorogenic acid.[48,66] Even if variation in this desirable character is often produced by differences in environmental and agronomic conditions, genetic variation also plays an important role in affecting the total amount and composition of flavonoids.

Table 1.1 Examples relating to patent applications for some new stevia varieties.

Patent number	Year	Country	Inventor/company	Varietal name/patent description	Characteristics
US 10564 P	1998	USA	Inventors: A.A. Marsolais, J. Brandle, E.A. Sys; registered by Royal-Sweet International Technologies Ltd.	RSIT 94-751	(1) High Reb A-to-Stev ratio; (2) high Reb A-to-Reb C ratio; (3) high Reb A-to-Dulc A ratio; (4) high percentage of SVglys consists of Reb A; (5) high concentration of Reb A; and (6) high sweetener concentration.
US 10562 P	1998	USA	Inventors: E.A. Sys, A.A. Marsolais; J. Brandle, registered by Royal-Sweet International Technologies Ltd	RSIT 94-1306	(1) High Stev-to-Reb C ratio; (2) high Stev-to-Reb A ratio; (3) high Stev-to-Dulc A ratio.
US 10563 P	1998	USA	Inventors: J. Brandle, E.A. Sys, A.A. Marsolais; registered by Royal Sweet International Tech.	RSIT 95-166-13	(1) High Reb C to Stev ratio; (2) high Reb C to Reb A ratio; (3) high Reb C to Dulc A ratio.
US 6031157 A (also published as US 6080561 A)	2000	USA	Inventors: T. Morita, Y. Bu; registered by Morita Kagaku Kogyo Co	Variety of *Stevia rebaudiana* Bertoni Morita SF5 and SF6	Reb A content 2.56 times more than Stev; this variety has a gene expressing a high content of Reb A and can dominantly transmit the gene to the next generation by seed propagation.
US 6255557 B1	2001	USA	Inventor: J. Brandle	*Stevia rebaudiana* with altered steviol glycoside composition	High level of total SVglys (about 14%) and a high ratio between Reb A and Stev (at least about 9:1).

(*continued*)

Table 1.1 (*continued*)

Patent number	Year	Country	Inventor/company	Varietal name/patent description	Characteristics
US 2011/0023192 A1	2011	USA	Inventors: T. Morita, K. Morita, S. Kanzaki	Novel stevia variety and method of producing sweetener	(1) Extremely-high content ratio of Stev; (2) this variety can be continuously produced from a seed; (3) method for preparing a sweetener which is extracted from dried leaves thereof; (4) a method for preparing α-glucosylstevioside from the sweetener.
US 7884265 B2	2011	USA	Inventors T. Morita, K. Morita, and K. Kornai; registered by Morita Kagaku Kogyo Co., Ltd.	High Rebaudioside A plant	(1) High content ratio of Reb A compared with Stev (at least 4 parts by weight or more of Reb A with respect to one part by weight of Stev); (2) a method for the production of a sweetener extracted from said plant and/or its dried leaves.
CA 2857089A1 (also published as, WO2012/088612 A1; US 20130347140; JP 2014502505 A)	2012	Canada	Inventor Q. Wang; registered by Glg Life Tech Corporation	High rebaudioside A plant and methods of producing the same and uses thereof	Method for breeding Stevia with a high content of Reb A; high yield of leaves; high content of total SVglys; high content of Reb A; strong resistance; stable traits of plants.

US PP23164 P3 (also published as US 20120090062)	2012	USA	Inventor Edgar Ramon Alvarez Britos; registered by PureCircle	AKH L1	Late harvest cycle; light green or yellow green leaves; high number of nodes on the main stem; high Reb A of total SVgly content (89%); high yielding of dried leaves at harvest.
US 20120090063 P1 (also published as US PP23728)	2012	USA	Inventor Edgar Ramon Alvarez Britos; registered by PureCircle	AKH L4	Early cycle; dark green leaves; medium number of nodes on the main stem; high number of basal buds; high Reb A content; high yielding of dried leaves.
WO 2014146084 A1 (also published as CN 105050388A, EP 2966990A1, EP 2966990A4, US 20160021918)	2014	USA	Inventors R. J. Brower, T. L. Carlson, B. Dang, M. D.Gonzales, M. Mc. Kennedy, N. E. Knutson; Registered by Cargill Incorporated	Stevia plants with an increased rebaudioside D content	High level of Reb D.
WO 2014/122227 (also published as CA 2899276A1, CN 105051195A, EP 2954058A2, US 2016/0186225, WO 2014/122227A3, WO 2014/122227A4, WO 2014/122227A8, WO 2014/122227A9)	2014	World Intellectual Property Organization	Inventors: M.D. Mikkelsen, J. Hansen, E. Simon, F. Brianza, A. Semmler, K. Olsson, S. Carlsen, L. Düring, A. Ouspenski, P. Hicks; registered by Evolva Sa	Methods for improved production of rebaudioside D and rebaudioside M	Methods for recombinant production of SVglys and compositions containing SVglys.

(*continued*)

Table 1.1 *(continued)*

Patent number	Year	Country	Inventor/company	Varietal name/patent description	Characteristics
WO 2016/094043 A1	2016	World Intellectual Property Organization	Inventors C.C. Shock, and C.A. Parris; registered by S&W Seed Company	SW 201	Excellent sweet leaf taste with very low bitterness and aftertaste; superior overwintering capability; high plant vigour; high leaf yield; late flowering; high SVgly levels.
WO 2016/085693 A1	2016	World Intellectual Property Organization	Inventors C.C. Shock, and C.A. Parris; registered by S&W Seed Company	SW 107	High productivity, including superior plant vigour and excellent overwintering capability; very sweet leaf taste with low bitterness and aftertaste and high Reb A content.
WO 2016/134449 Al	2016	World Intellectual Property Organization	Inventors Q. Wang, Y.L. Zhang, and C.K. Li	High rebaudioside a plant varietal, methods of extraction and purification therefrom, of compositions with enhanced rebaudioside a content and uses of said composition	Reb A content of 6–20% dry weight and total SVgly content of 15–28% dry weight in the leaf. The plant is developed using selective breeding technologies and identified by RAPD gene analysis.

1.3 Variability of Steviol Glycosides Content and Composition in Stevia Leaves

1.3.1 Source of Variation

Caligari[67] stated "*the breeder generally uses the natural variation that already exists within the species and he needs to measure any character in order to observe variation in its expression*". Often, this variation reflects not just variation produced by differences in the environment in which the plant grows, but also genetic variation, which is heritable. Sources of variation, in fact, include environmental variation, genetic variation and interaction of genetic and environmental variation. Generally, plant breeders must distinguish among these sources of variation for the character of interest in order to effectively select and transmit the desired character to subsequent generations. Consequently, in order to effectively manipulate quantitative and qualitative traits of interest in stevia, the breeders need to understand the nature and the extent of their genetic and environmental control. What the breeder really observes is the phenotype of the plant, but he needs to select the genotype.[67] The success of a breeding program in stevia depends on the selection of potential plants with desirable characteristics from the populations, and, therefore, on the ability to predict the genotypic value of the selected plants, starting from the phenotypic value or from some other selectable criterion.[68] The phenotype of a plant is modelled as a function of its genotype as modified by the environment.

Some characters are more responsible or sensitive to growing conditions than others. It is known that in stevia, SVglys accumulation and composition, widely vary according to phenological stage and growth conditions (photoperiod, irradiance, temperature, and available nutrients),[8,9,16,26,65,69–77] as well as according to agronomic practices adopted during cultivation (such as fertilisation, water availability and salinity, plant density, harvest time and frequency).[9,18,19,31,44,78–84]

The degree of sensitivity and the potential plant response to the environment is determined by the genetic composition of the individual plant or population of plants. So, genetic variability is essential in order to make progress in stevia cultivar improvement.

1.3.2 Phenotypic and Genetic Variation in *S. rebaudiana*: The Genetic Control of SVgly Composition

Today, there are about 90 varieties of *S. rebaudiana* developed throughout the world.[16,40,85] The most known and studied stevia varieties are Criolla and Morita II; the first seems to be the original stevia variety native to Paraguay, while the second variety has been selected for its high Reb A content. Other known and cultivated varieties are Eirete, a hybrid developed in Paraguay for intensive cultivation; Morita III, obtained from Morita II and characterised by low water requirements; and Katupyry, a recently selected variety in Paraguay for growing in arid soils, characterised by high sweetening power. Also, several new stevia varieties with improved desirable traits have been released, such as AKH L1; AKH L4; SW 201; SW 107, and others (see Table 1.1).

The genetic divergence among stevia genotypes plays an important role in the selection of parents having wider variability for different traits. Consequently, the study of phenotypic and genotypic diversity plays a key role in order to develop a plant breeding program for stevia.[6] In stevia, considerable genetic and phenotypic variability has been observed, with particular regard to plant size, flowering period, leaf yield and SVgly content and composition.[5,20,86–89]

In Figure 1.1, an example of phenotypic variability between two stevia genotypes, belonging to the germplasm collection of the Department of Agriculture, Food and Environment of the University of Pisa, is reported. This natural variability could be partially due to the largely self-incompatible nature of the flowers, as reported by Handro *et al.*[90] and Midmore and Rank.[91]

As the varietal diversity in stevia is reportedly quite high, researchers have attempted to enhance their knowledge about this variability and to develop a breeding strategy, studying the genotypic and phenotypic variability of several genotypes under different environmental and agro-climatic conditions.

Since a better understanding of the genotypic control of SVgly composition could be of interest for plant breeding, initial investigations on the genotypic variability for SVgly composition within a population have established three clusters: (1) plants containing mainly glucose(glc)-type glycosides (Stev and Reb A); (2) plants containing mainly rhamnose(rhm)-type glycosides

Figure 1.1 Phenotypic variability between two *S. rebaudiana* genotypes, in trials carried out at the Experimental Centre of the Department of Agriculture, Food and Environment, of the University of Pisa.

(Reb C, Dulc A) and (3) plants containing nearly equivalent quantities of glc- and rhm-type glycosides.[92] As a result of differential glycosylation, each SVgly has distinctive organoleptic and biological properties. It is known that a sugar unit or a carboxyl group in C19 position and a sugar or a hydroxyl group in C13 position, are essential for sweetness.[93] However, rhamnosylation decreases the organoleptic properties, and the resulting sweetness and the taste quality of the rhamnosylated SVglys (such as Dulc A and Reb C) is inferior to their glycosylated counterparts.[94]

Brandle and Rosa[43] showed that economically important breeding traits of stevia are characterised with high variability within populations and high heritability. These traits include leaf yield (h^2 = 62.1), leaf-to-stem ratio (h^2 = 78.8) and content of SVglys (h^2 = 76.6). Due to high heritability they are susceptible to modification by selection.[43] Brandle[49] studied the genetic control of the proportions of Reb A and Reb C, suggesting that these two compounds, equally glycosylated, differing only in the nature of one sugar unit, are synthesised by the same enzyme. These findings have been recently confirmed by Barbet-Massin *et al.*[53] in a study aimed to evaluate the genotypic variability for SVgly content and composition in several stevia genotypes. In this two year experiment, a set of 96 stevia genotypes was randomly chosen in a population of Criolla, reproduced by stem cuttings and then transplanted in open field conditions, at the INP-EI Purpan Nursery Station, Toulouse, France. Five of these 96 genotypes were also transplanted at the INP-EI Purpan Experimental Station, Seysses, France. The genotypes showed a high variability in SVgly content and composition, with higher Reb A than Stev content in some genotypes, whereas other genotypes had a very high proportion of minor SVglys. In addition, while SVgly content varied among the different environments here tested, SVgly composition remained stable, indicating a higher genotypic determinism for this trait than for total content. Lankes *et al.*[95] studied the performance of six stevia genotypes in temperate European climatic conditions, in comparison with Gawi, a genotype selected by the University of Bonn. The six genotypes were provided by the European Stevia Association (EUSTAS). The results showed that Gawi was well adapted to the environmental conditions of European temperate zones, but genotype C was able to outperform Gawi in terms of leaf yield and Reb A content. Anami *et al.*[96] evaluated 15 stevia clones (selected for their high concentration of total SVglys and considerable Reb A to Stev ratio) and characterised genetic divergence in order to select genitors for hybridisation in breeding programs. Clones were evaluated in field experiments in the North region of Paraná, adopting a randomised complete blocks scheme with three repetitions. These authors observed an ample genetic variability among the 15 stevia clones for plant height, fresh and dry matter production, number of branches per plant, SVgly concentration and Reb A/Stev ratio, and they concluded that the Reb A/Stev ratio and total SVgly concentration contributed the most for genetic divergence (79% and 12%, respectively). Four of the 15 clones showed high mean genetic divergence in relation to the whole genotypic pool studied. Therefore, the authors stated that hybridisations among these four clones could provide a generation of segregated populations with high genetic potential for superior individuals' selection.[96]

1.3.3 Character Association and Heritability in *S. rebaudiana*

For selection and breeding purposes in stevia, information about the interrelationships among characters are of crucial importance for the following reasons: (1) to have a good choice of characters for selection of desirable genotypes under planned breeding program for higher yield, yield components and qualitative traits; (2) to have information about the knowledge of nature and magnitude of variation existing in available breeding materials and (3) to know the association among characters. In stevia, high estimates of heritability, genotypic and phenotypic variability were recorded for several important characters. Several studies have been carried out in stevia in order to find character associations potentially useful in assisting selection programs and to demonstrate the dependence of yield on various growth parameters, as well as Stev content.[2–4,43,97–99]

First reports[43] demonstrated the presence of significant heritable variation for some important characters and underlined the possibility of the genetic improvement for this species. In particular, Brandle and Rosa,[43] studying a landrace stevia cultivar (imported from China and grown in Delhi Research Station, Ontario, Canada), found that there was high heritability for three economically important characters: leaf yield, leaf–stem ratio, and Stev concentration. Furthermore, the authors observed that Stev concentrations were uncorrelated to yield or leaf–stem ratio, indicating that concurrent improvement of agronomic and chemical characteristics was possible. Several authors[2–4] observed that, in stevia, plant leaf yield was proportional to branch number, leaf number and (not always) plant height. Tateo *et al.*[100] found that total Stev content was positively correlated with the leaf–stem ratio, while Shyu *et al.*[99] concluded that leaf thickness and Reb A/Stev ratio were positively correlated. High Reb A content was found to be also linked to large leaf area, high net photosynthetic rate, high chlorophyll and protein content.[101] Nakamura and Tamura[92] found positive correlations between Dulc A and Stev, as well as between Reb A and C; on the contrary, negative correlations have been observed between Stev and Reb A and between Dulc A and Reb C.

1.4 The Chromosome Number in Stevia Genus

The genus stevia shows great variation in chromosome number.[5,35] Although most reports indicate that $n = 11$ ($2n = 22$),[86,89,102,103] values of $2n = 24, 33, 34, 44, 48, 66, 70$ have also been observed.[104–116] Among these, strains with $2n = 33$ and $2n = 44$ (representing triploid and tetraploid cytotypes) showed a high degree of male sterility owing to the chromosomal abnormalities during gamete formation.[4] The first and extensive cytological study on many species of the genus stevia (belonging to the following Series: *Corymbosae* 16 species counted out of 38, *Fruticosae* 9/25; *Podocephalae* 5/16) was performed by Grashoff *et al.*[117] These authors found that, in North America, the shrubby species presented a gametic chromosome number of n = 12; the herbaceous species with flower heads in a lax paniculate cluster had a number of n = 11 (with no aneuploidy), while species in compact corymbose cluster had $2n$ = 34 univalents with aneuploidy.[118] This latter group of species can be considered a triploid derivative of n = 11,[119] due to the widespread occurrence of apomixis among the North American species. For South American species, Grashoff *et al.*[117] established that the species were diploid with the exception of hexaploid *S. elatior* ($2n$ = 66) from Colombia.[120] The studies carried out on this topic indicate a clear predominance of the basic chromosome number x = 11 among stevia species from South America, with only three species (*S. lucida* from Colombia, one population of *S. jujuyensis* from Argentina, and *S. organensis* from Brazil) showing x = 12.[5,121–123] As reported by Frederico *et al.*,[102] these three species may have originated by ascending aneuploidy from species with x = 11. This implies that the main mechanism in the evolution of the South American species of stevia is probably chromosome inversions, with a small amount of aneuploidy and polyploidy.[5,102] Regarding to chromosomal morphology, Frederico *et al.*[102] studied the karyotypes of six stevia species from southern Brazil, utilising root tip metaphase. These authors found that all the species analysed were diploid, with a chromosome number of $2n$ = 22, with the basic chromosome number of x = 11. The chromosomes of the karyotypes were also similar in size (1.0–2.4 µm) and most of them were metacentric, with a variable number of submedian ones. Only *S. ophryophylla* and *S. rebaudiana* had a pair with a subterminal centromere.[102] Frederico *et al.*[102] suggested the occurrence of pericentric inversion as the rule in the divergence of Brazilian species of stevia. More recently, de Oliveira *et al.*[89] determined the chromosome number, the meiotic behaviour and morphological features, in diploid and polyploid cytotypes of *S. rebaudiana*. In this study, tetrad analysis and pollen viability as well as stomata size, pollen diameter, and some vegetative (leaf size and plant height) and reproductive (inflorescences per plant) features, were also assessed. These authors[89] confirmed for *S. rebaudiana* the chromosome number of $2n$ = 22 (n = 11), previously reported by Frederico *et al.*[102] and Monteiro,[86,103] with the exception of two strains, characterised $2n$ = 33 and $2n$ = 44, and representing triploid and tetraploid cytotypes, respectively. de Oliveira *et al.*[89] observed for *S. rebaudiana* the presence of a nucleolus organising region on the short arm of the third major chromosome pair, confirming previous reports.[102]

1.5 Plant Breeding Programs in Stevia

Plant breeders employ a variety of techniques to improve genetic composition of the crops and different breeding approaches are required for self-pollinating, cross-pollinating and clonally-propagated plants. In cross-pollinated species, like stevia, the selection of new cultivars involves the change of the gene frequency of desirable alleles within a population of mixed genotypes, trying to maintain a high degree of heterozygosity.[67]

At the beginning, the development of new varieties of stevia with a higher content of Reb A and a reduced content of Stev was the primary aim of plant breeders.[43,49,52–54,95,96,124–126] At the same time, also the morphological characteristics, such as plant height, branches and leaf number, leaf area index have been taken into account, together with the photoperiodic needs of stevia for its vegetative development, in order to ensure optimal production of SVglys in the different latitudes. Other characteristics such as tolerance or resistance to disease and pests, or to drought or high soil pH values should be further considered.[5]

There are several methods for genetic improvements and the most breeding programs, for stevia, are based on conventional plant breeding approaches such as selection and intercrossing among various desirable genotypes, along with chemical profiling for improving quality traits. Since all traits of a plant are controlled by genes located on chromosomes, conventional plant breeding can be considered as the manipulation of the combination of chromosomes, through pure line selection, hybridisation and polyploidy (increased number of chromosome sets).

1.5.1 Selection Programs in Stevia

Mass selection is a very simple breeding scheme, that allows to create a new population by cross-pollinating two different existing open-pollinating populations.[127] A representative set of individuals from each population will be taken to be crossed. The seed that results from such a set of crosses is grown under field conditions over a number of seasons.[128]

As reported by Yadav *et al.*,[5] three decades of breeding and selection programs in stevia have allowed to increase SVgly concentration in the leaves by as much as 20%.[45] At the beginning, the countries that have been researching more in stevia variety improvement were Japan, China, Korea, and Taiwan, reporting success in their breeding/selection programs with the releases of new varieties with enhanced SVgly content and higher leaf yields.[5,129] Some examples of these varieties/cultivar selections and releases are:

- Sunweon 2, characterised by high leaf yield and Stev content[130]
- Sunweon 11, with a high percentage of Reb A[52]
- K1, K2 and K3, characterised by high crop yield and favourable RebA–Stev ratio[131]
- Zongping, with increased Reb A and Stev content in the leaves[131]
- SM4, with high yield and RebA–Stev ratio.[101]

Some of these selections, although very high yielding, are self-incompatible and can only be reproduced vegetatively.[52] This limits their commercial use, although they may be useful for breeding new hybrids.

Patent applications for superior stevia plant types, in particular with improved sweetening power and taste, have been filed and registered (especially in USA and Japan) as results of selection, intercrossing and hybridisation.[5] In the US and the EU, the number of patent applications regarding new stevia varieties and SVglys, strongly increased after the JECFA evaluation on the safety of SVglys in 2008. Over 1000 patent applications concerning stevia had been filed by the end of 2014. Based on Espacenet data, eight companies (PureCircle®, Pepsi Co, Coca-Cola®, DSM, Evolva, McNeil Nutritionals LLC, Suntory Holdings and Cargill®) have made the most patent applications focused on SVgly production methods.[132] With their patents, these companies are able to control the whole market for SVglys.[132]

Examples of patent applications (from 1998) for several new stevia varieties have been reported and are shown in Table 1.1.

The well-known stevia varieties obtained through selection and intercrossing are RSIT 94-1306,[133] RSIT 94-751,[134] and RSIT 95-166-13[26] (Table 1.1). The first two were obtained from seeds of a "landrace" variety, collected from China and are characterised by improved content of SVglys. RSIT 95-166-13 was developed as a unique combination of characteristics, which distinguished it from its parents and all other stevia varieties due to its high and unique Reb C–Stev ratio. Due to stevia self-incompatibility, these varieties cannot be reproduced using a seed-based production system, and they have to be vegetatively propagated by shoot-tip and stem cuttings.[5]

The selection was based on phenotypic traits, which depend very much on the breeding conditions (*i.e.* environmental conditions, such as soil and weather) and on the plant age.[135] It is estimated that, in the case of young seedlings, only 20–30% of the variability is genetically determined.[5,135] Therefore, selection should be made with mature plants, which means a longer time before obtaining the results.[135]

1.5.2 Recurrent Selection

In order to improve quantitatively inherited characteristics in stevia—a cross-pollinated species characterised by high differentiation of a given trait within the population—recurrent phenotypic selection tends to be more effective than mass selection.[54,127,135]

This technique systematically increases the frequency of favourable alleles and maintains the genetic variation within a population to permit continual progress from selection. Most methods of recurrent selection include three phases conducted in a repetitive manner: development of progenies, evaluation of progenies in replicated trials and recombination of the superior progenies based on the evaluation trials.[136]

1.5.3 Synthetic Cultivars

A synthetic cultivar refers to a cultivar produced by intercrossing clones or sibbed lines obtained from a breeding population during several cycles of recurrent selection. Caligari[67] stated that "*a synthetic cultivar basically gives rise to the same end result as an open-pollinated cultivar, the main difference being that a synthetic cultivar is continually reproduced from specific parents, whereas if it is left to open-pollinate to produce over generations, it will change its genetic make-up as a population*". A synthetic cultivar is propagated for only a limited number of generations and then must be reconstituted from the parental stock.

In the case of stevia, the development of synthetic varieties appears to be the most suitable method for the development of a variety enriched in Reb A and having a high SVgly content.[5] A synthetic variety, 'AC Black Bird', has been developed by Brandle.[137] This cultivar is characterised by a total SVgly content ranging from 14.0–18.8% with a high Reb A–Stev ratio of at least 9.1 : 1 and up to 11.0 : 1.[137] In order to achieve the parents for the synthetic cultivar, crosses were made among a high number of progeny plants, grown in the field, and the selection was made among these plants. Seeds were collected and a large number of progeny (12 families and 60 plants for each family) were obtained and planted in the field. Twenty plants from each family were selected by HPLC analysis for their SVgly content. Then, four of these plants were further selected on the basis of their SVgly content and composition (low Stev, high Reb A, and high Reb A-to-Stev ratio). These four selected plants were enclosed and cross pollinated by bees, and, at the end of the reproductive phase, their seeds were mixed together, and the new progeny obtained.

Another synthetic stevia cultivar was also developed by Morita and Yucheng,[56] the ATCC Accession No. PTA-444. These breeders, through repetitive crossing and selection, obtained a variety with at least 2.56 times more Reb A than Stev and capable of being cultivated by seed propagation, differently to "AC Black Bird" variety, that is self-sterile and can be only bred vegetatively.[135]

1.5.4 Changing the Number of Chromosomes in Plant Cells

1.5.4.1 Polyploid Breeding

Manipulation of ploidy is considered as a valuable tool in breeding programs of many plants,[138] in order to increase the organ size, and the adaptability of individuals and to improve the agronomic yield.[139,140] Polyploidy often generates variants that may possess useful characteristic by doubling the gene products.[138]

First studies showed that triploid stevia plants (3*n*)—produced by mating tetraploid plants (4*n*) with normal diploid plants (2*n*)—were characterised by higher Reb A content and larger leaves than normal diploid plants.[141,142] However, Shaopan *et al.*,[143] Yadav *et al.*[140] and de Oliveira *et al.*[89] revealed that tetraploids in stevia determine larger leaf size, thickness and chlorophyll content with increased biomass in comparison with diploid strains. These findings underline that tetraploid stevia genotypes can represent important germplasm for further improvement in terms of biomass yield and SVgly content. Rameshsing *et al.*[139] developed different stevia mutants using colchicine at different concentrations and confirmed that colchicine represents an effective polyploidising agent in creating new stevia variants with improved biomass and SVglys. Hedge *et al.*[144] recorded higher numbers of secondary branches, greater leaf thickness and area, delayed flowering and higher SVgly content in polyploid stevia plants in comparison with normal diploids. In addition, these authors found that yield performances of all induced polyploids were better compare to diploids and, among the polyploids, mixoploids showed the best yielding ability. de Oliveira *et al.*[89] studied the pollen viability of triploid and tetraploid strains and observed that all polyploids had non-functional pollen. In particular, these authors observed that the higher the ploidy number, the greater the size of the pollen and the stomata and the lower their number per unit area. In addition, they found that the triploid strain produced the shortest plants and the lowest number of inflorescences, whereas the tetraploid strain had the largest leaves.[89]

1.5.4.2 Anther Culture

Anther culture, usually carried out at the beginning of breeding program, leads to haploid plants from which doubled haploids/homozygous plants can be developed by colchicine treatment in a short time.[5] Through anther culture, by developing haploid progenies, dominant and recessive characters can be separated and varieties with novel characters can be developed. This method is relevant for stevia, since development of a pure line in stevia is very problematic due to its self-incompatibility. Through anther culture, pure-line from a single stevia plant can be obtained. Flachsland *et al.*[145] successfully carried out stevia plant regeneration from anthers, even if cytological studies of root tips from regenerated plants have been revealed a normal diploid number of chromosomes ($2n = 22$). Garnighian[146] developed a patent for the stevia variety "T60", starting from anther culture of the Paraguayan variety "Criolla", to produce a haploid line with enhanced SVgly content. This haploid line had undergone a colchicine treatment in order to obtain diploid fertile plants.

1.5.5 Marker-assisted Selection

Marker-assisted selection (MAS) is a procedure that has been developed to avoid problems associated with phenotypic selection, replacing the selection of the phenotype by selection of genes, both directly and indirectly.[147] In fact, molecular markers are not influenced by the environment and are detected at any stage of plant development.[148] DNA markers are used to assess genetic diversity at various levels of taxon-species, inter and intra population and progeny. One of the most efficient molecular methods, in terms of ability to produce abundant polymorphic markers within a short time and at low cost, is the random amplified polymorphic DNA (RAPD) technique. Recently, RAPD has been used for estimation of genetic diversity in various plant species, collected from different geographical regions.[149] The first genetic linkage map for *S. rebaudiana* was constructed by Yao *et al.*[150] based on RAPD markers using segregation data from a pseudo test-cross F1 population. This information will be useful to those interested in developing marker-assisted selection procedures and quantitative trait analysis, as well as provide a starting point for those interested in genome organisation in stevia. Recently, Chester *et al.*[151] evaluated the genetic and metabolic variability in *S. rebaudiana* among accessions of different geographical regions of India, using RAPD markers and high performance thin layer chromatography (HPTLC) analysis. Similarly, Thiyagarajan and Venkatachalam,[149] while investigating the genomic DNA polymorphism and phytochemical variation in stevia by RAPD-PCR and HPLC analysis, concluded that there was a strong correlation between the phytochemical variables and the DNA polymorphism data. The correlation between RAPD markers and chemotype highlights the potentiality of RAPD analysis as a reliable method for identification and authentication of high yield accessions of SVgly production.

With the aim to develop a breeding strategy to get cultivars adapted to organic production in South West of France, Hastoy *et al.*[125] evaluated the genetic variability using SSR (Simple Sequence Repeat) molecular markers in five genotypes from different origins, cultivated under greenhouse conditions, as well as well as in 33 additional genotypes whose leaves were harvested in the field. Molecular markers have been developed in order to classify the genetic diversity of the available stevia germplasm, selecting the SSR markers on the basis of the available sequences present in database. This preliminary study revealed a very great phenotypic and genetic variability in the tested genotypes. The authors concluded that the different genotypes could be classified through their genetic distance and the five SSR markers used in this experiment were sufficient to distinguish all the genotypes and allow partial classification of the genotypes into five different clusters.

1.5.6 New Biotechnological Tools

New biotechnological tools, such as random mutagenesis, site direct mutagenesis, transgenic plants, *Agrobacterium* mediated gene transfer and silencing, can provide useful opportunities for improvement of desired traits in stevia. These techniques can bring innovations beyond the usual approaches, with consequent technical and scientific progress in plant breeding.

1.5.6.1 Random Mutagenesis

Induced mutation has been considered an important phenomenon which, when combined with *in vitro* studies, could act as a very handy tool for creating novel genetic variations with desirable traits.[152] In order to obtain this variance towards desirable genetic traits by means of DNA alteration, it is possible to use chemical (ethyl methanesulfonate, EMS; *N*-methyl-*N*′-nitro-*N*-nitrosoguanidine, NTG) or physical (gamma radiation, hard X-ray, neutron beam) agents. In any case, random mutagenesis by chemical or physical agents is a blind technique, and outcomes are often unpredictable. In fact, the induced mutation can deactivate or silence genes with some specific biological activity. The seeds of stevia exposed to gamma irradiation do not show changes in germination percentage, even if at very high doses root development can be suppressed.[5] Toruan-Mathius *et al.*[153] have used Cobalt-60 γ-rays irradiation to induce variation in stevia breeding lines. The authors produced plants with the highest growth rate, normal morphology and highest Stev content by using gamma radiation at the dose of 1500 rad. Khan *et al.*[154] exposed stevia leaf explants to different doses of EMS and γ-ray. The results showed lower chlorophyll content, transpiration rate, CO_2 exchange rate and stomatal conductance in traded plants than in control (plants without any chemical or physical treatment). However, at harvesting time, phytochemical analysis showed that plants treated with γ-rays had double Reb A content with respect to the control plants, and the EMS-treated plants showed a 1.5- to 2.0-fold increase of both Stev and Reb A, in comparison with control plants. Khan *et al.*[154] found that SVgly enhancement was related to an increasing UGT74G1 (corresponding to Stev biosynthesis) and UGT76G1 (corresponding to Reb A biosynthesis) gene expression, in both EMS- and γ-ray-treated plants.

1.5.6.2 *Site-directed Mutagenesis*

Site-directed mutagenesis is a molecular biological method and can be applied to any given plant for creating a specific mutation in a known sequence. This method makes largely use of the Expressed Sequence Tags (ESTs), which are contributing to the gene discovery in plant secondary metabolism, revealing gene expression patterns, gene regulation and sequence diversity.[155] Some studies were conducted about stevia ESTs, in order to collect a resource for gene discovery, understand and increase SVgly biosynthesis.[155–157] In particular, Brandle *et al.*[155] sequenced 5548 ESTs from a *S. rebaudiana* leaf cDNA library. In this study, ESTs involved in diterpene synthesis were identified underlining that the use of ESTs greatly facilitated the identification of candidate genes and increased the understanding of diterpene metabolism. Brandle *et al.*[155] grouped ESTs into functional categories and found that significant portions of the ESTs were specific for standard leaf metabolic pathways; energy and primary metabolism represented 17.6% and 13.1% of total transcripts, respectively. Regarding ‘secondary metabolism’, 278 ESTs were classified into this category and, among these ESTs, 62 were candidates for diterpene glycoside synthesis. Diterpene metabolism in stevia represented 1.1% of total transcripts. This study identified candidate genes for 70% of the known steps in the SVgly pathway, with kaurene oxidase the eighth most abundant EST in the collection.

1.5.6.3 Transgenic Plants

Stevia is a cross-pollinated species that requires the presence of insect pollinators to produce viable seeds.[38,158] In order to obtain homogeneous population by selected plants, *in-vitro* propagation appears to be the best approach, even if it is not economically sustainable for large-scale production, because of its high costs and high labour inputs. Therefore, the possibility to have alternative methods for rapid production of homogeneous stevia populations, characterised by high and uniform yield of SVglys, is strongly needed. Therefore, transgenic engineering may represent an interesting resource. Despite the variety of genetic engineering techniques, few studies have been carried out on stevia. Mubarak *et al.*[159] produced transgenic stevia plants using biolistic gene gun protocol; carrying *bar* gene, as a selectable marker for herbicide bialaphos resistance[160] and GUS gene (*Escherichia coli* beta-glucuronidase gene), as a reporter marker for analysis of gene expression in transformed plants.[161] The *bar* gene was originally cloned from *Streptomyces hygroscopicus*, an organism that produces the tripeptide bialaphos as a secondary metabolite. In the study of Mubarak *et al.*,[159] histochemical staining technique and PCR analysis were used to detect the presence of the GUS and *bar* genes in the putatively transformed tissues. The expression of GUS DNA, introduced into the plant cells, was monitored by using x-gluc (5-bromo-4-chloro-3-indolyl glucuronide) as a substrate for the GUS enzyme. The results demonstrated that the transformed *S. rebaudiana* explants were more resistant than non-transformed one, with 66.7% of transformed stevia plants able to survive at 3 mg L^{-1} of bialaphos.

1.5.6.4 Agrobacterium Mediated Gene Transfer

The ability of *Agrobacterium tumefaciens* to transfer DNA to plant cells has been utilised in plant genetic engineering. *Agrobacterium*-mediated plant transformation is a highly complex and evolved process involving genetic determinants of both the bacterium and the host plant cell.[162] *A. tumefaciens* is capable to transfer a particular DNA segment (T-DNA) of the tumour-inducing (Ti) plasmid into the nucleus of infected cells, where it is subsequently stable integrated into the host genome and transcribed causing the crown gall disease.[163,164]

Khan *et al.*[165] reported *Agrobacterium tumefaciens* (EHA-105 harboring pCAMBIA 1304)-mediated transgenic plant production *via* direct regeneration from leaf and elite somaclones generation in *S. rebaudiana*. The development of somaclones *via* indirect regeneration was also evaluated to obtain varieties with improved traits, *e.g.* disease resistance, improved tolerance to climatic condition, and efficient yield of desirable phytomolecules.[165] In this experiment, transgenic plants were compared with four different somaclonal plants and mother plants through *in vitro* leaf culture. These authors expected that the random insertion of T-DNA into plant genome could affect the total SVgly content with the consequent possibility to develop an improved variety of stevia plants. The obtained results showed that inter-simple sequence repeat (ISSR) profiling of transgenic and somaclonal plants showed a total of 113 bands, out of which 43% were monomorphic and 57% were polymorphic. Transgenic plant was found to be closer to mother plant, while, on the basis of steviol, Stev, and Reb A profile, somaclone S2 was found to be the best (with maximum Stev content) and maximum variability in ISSR profiling.

1.5.6.5 *Agrobacterium Mediated Transient Gene Silencing (AMTS)*

Gene silencing has always been a recommended tool to understand the functionality of genes.[166,167] RNA interference (RNAi) being the most commonly employed approach that involves preparation of inverted repeat construct of target gene and generation of transgenics.[167] *Agrobacterium tumefaciens* is being used as an efficient molecular vehicle to transform plants with silencing/overexpression constructs for transgenic development.[168] Guleria and Yadav[166] adopted a gene silencing approach in order to understand the genetic regulation of SVgly biosynthetic pathway in stevia plant, and its molecular mechanism and association with gibberellins. RNA interference (RNAi) was applied through AMTS. It is known that SVglys share common steps of their biosynthesis with the gibberellin pathway.[41,146] However, the genetic basis of SVgly biosynthesis and its relation with gibberellins is not well understood.[166] In the SVgly pathway, steps catalysed by enzymes encoded by SrKA13H, SrUGT85C2, SrUGT74G1 and SrUGT76G1 genes have been considered important for SVgly biosynthesis.[169] On the basis of these considerations, Guleria and Yadav[166] silenced SrKA13H and three SrUGTs (SrUGT85C2, SrUGT74G1 and SrUGT76G1) genes encoding ent-kaurenoic acid-13 hydroxylase and three UDP glycosyltransferases of SVgly biosynthesis pathway. The downregulation of the above-mentioned genes was found to negate the SVgly accumulation level. However, the total SVgly reduction was not more than 60% on AMTS of any of the four genes, suggesting the existence of alternate SVgly biosynthesis route. In particular, SrKA13H and SrUGT85C2 were identified as regulatory genes influencing carbon flux between SVgly and gibberellin biosynthesis. In fact, silencing of these two genes was found to block the metabolite flow of SVglys.

1.6 Conclusions

The worldwide demand for high potency and natural non-calorific sweeteners is expected to increase in the near future, which would necessitate new stevia cultivars with high quality yields of sweet compounds.

Great progress has been made so far in stevia breeding strategies. The data presented in this chapter clearly indicates the growing interest of scientists, researchers and plant breeders in *S. rebaudiana* genetic improvement through conventional and innovative biotechnological approaches. Stevia plant breeding will continue to be highly dependent on classical techniques but the adoption of new biotechnological approaches that will be used in parallel with the more classical ones will contribute to increase the efficiency and effectiveness of breeding efforts in this species.

However, despite the progress that has been made in stevia breeding strategies, some critical questions should be taken into consideration in the future research. First of all, a number of concerns have arisen due to the application of plant transformation technology with particular regards to new transgenic crops. Therefore, researchers and plant breeders should take into consideration these concerns and the regulations applied to plants obtained using recombinant DNA.

Another urgent question regards patents of new stevia varieties. Stevia varieties together with SVglys are the subject of intense patent activity. The International Union for the Protection of New Varieties of Plants (UPOV) database showed that there are some 40 applications worldwide for Plant Breeders Rights (31) or Plant Patents (9) concerning stevia.[170] Regarding patent applications for new stevia varieties, there is the issue with intellectual property of the indigenous Guaraní people in Paraguay and Brazil. In fact, since global demand for natural and sugar-free products is rapidly expanding, stevia plants are being grown and processed commercially in many countries outside Paraguay, especially in China. However, the Guaraní people's right to benefit from its use, as established under the Convention on Biological Diversity's Nagoya Protocol, is being ignored and is a clear case of biopiracy.[132]

References

1. L. G. Angelini, A. Martini, B. Passera and S. Tavarini, in *Reference Series in Phytochemistry. Sweeteners, Pharmacology, Biotechnology, and Applications*, ed. J.-M. Mérillon and K. G. Ramawat, Springer Reference, 2016, DOI: 10.1007/978-3-319-26478-3 8-1.
2. L. Buana and D. H. Goenadi, *Menara Perkebunan*, 1985, **53**(3), 68.
3. L. Buana, *Menara Perkebunan*, 1989, **57**(2), 29.
4. S. Z. Shu and W. Z. Wang, *Acta Agron. Sin.*, 1988, **14**, 167.
5. A. K. Yadav, S. Singh, D. Dhyani and P. S. Ahuja, *Can. J. Plant Sci.*, 2011, **91**, 1.
6. H. S. Othman, M. Osman and Z. Zainuddin, *Aust. J. Basic Appl. Sci.*, 2015, **9**, 1.
7. N. I. Bondarev, M. A. Sukhamova, O. V. Reshetnyak and A. M. Nosov, *Biol. Plant.*, 2003, **47**, 261.
8. S. Ceunen and J. M. C. Geuns, *Plant Sci.*, 2013, **198**, 72.
9. S. Tavarini, I. Pagano, L. Guidi and L. G. Angelini, *Plant Biosyst.*, 2016, **150**, 953.
10. M. Alvarez, *Stevia rebaudiana Bert. Estado Atual Do Conhecimento*, Universidade Estadual de Maringa, Maringa, 1984, p. 118.

11. A. Bertonha, A. S. Muniz, J. W. P. Carneiro, E. N. Martins, I. C. Jabur and S. I. Thomaz, in *Estudo de cultivo, reproducao e selecao das variedades mais productivas de Stevia rebaudiana, en solos do norte de Prana*, UEM, Mimeo, Maringa, 2nd edn, 1984, p. 103.
12. C. C. Shock, *Agron. Prog. Rep.*, 1982, 122.
13. R. Monteiro, *Taxonomia e bologia da reproducao de Stevia rebaudiana. Campinas. Dissertacan (Mestrado)*, Instituto de Biologia, Universidade Estadual de Maringa, 1986, p. 104.
14. O. Katayama, T. Sumida, H. Hayashi and H. Mitsuhashi, *The Practical Application of Stevia and Research and Development Data*, I.S.U. Company, Japan, 1976, p. 747.
15. T. Sumida, *Misc. Publ. Hokkaido Natl. Exp. Stn.*, 1973, **2**, 79.
16. K. Ramesh, V. Singh and N. W. Megeji, *Adv. Agron.*, 2006, **89**, 137.
17. E. W. Mass and G. J. Hoffman, *ASCE J. Irrig. Drain. Div.*, 1977, **103**, 115.
18. Y. Zeng, L. P. Guo, B. D. Chen, Z. P. Hao, J. Y. Wang, L. Q. Huang and M. L. Chen, *Mycorrhiza*, 2013, **23**, 253.
19. M. Reis, L. Coelho, G. Santos, U. Kienle and J. Beltrão, *Agric. Water Manage.*, 2015, **152**, 217.
20. N. W. Megeji, J. K. Kumar, S. Virendra, V. K. Kaul and P. S. Ahuja, *Curr. Sci.*, 2005, **88**, 801.
21. C. Ishiba, T. Yokoyama and T. Tani, *Ann. Phytopathol. Soc. Jpn.*, 1982, **48**, 44.
22. N. M. Lovering and R. D. Reeleeder, *Plant Dis.*, 1996, **80**, 959.
23. K. F. Chang, R. J. Howard and R. G. Gaudiel, *Plant Dis.*, 1997, **81**, 311.
24. S. C. Thomas, in *Medicinal Plants: Culture, Utilization and Phytopharmacology*, Technomic Publishing Co. Inc., Basel, Switzerland, 2000, p. 517.
25. A. Kamalakannan, V. Valluvaparidasan, K. Chitra, E. Rajeswari, K. Salah Eddin, D. Ladhalakshmi and A. Chandrasekaran, *Plant Pathol.*, 2007, **56**, 350.
26. J. E. Brandle, A. N. Starratt and M. Gijzen, *Can. J. Plant Sci.*, 1998, **78**, 527.
27. S. Bhandari and N. S. K. Harsh, *Indian For.*, 2006, **132**, 385.
28. L. G. Angelini and S. Tavarini, in *Final Report Eridania Sadam Project Contract*, 2010, p. 80.
29. K. H. Kang and E. W. Lee, *Korean J. Crop Sci.*, 1981, **26**, 69.
30. A. D. Kinghorn, in *Stevia, the Genus of Stevia, Medicinal and Aromatic Plants-industrial Profiles*, ed. A. D. Kinghorn, Taylor and Francis, London, 2002, pp. 1–17, ISBN 0-415-26830-3.
31. M. Serfaty, M. Ibdah, R. Fischer, D. Chaimovitsh, Y. Saranga and N. Dudai, *Ind. Crops Prod.*, 2013, **50**, 731.
32. L. B. P. Zaidan, S. M. Dietrich and G. M. Felippe, *Jpn. J. Crop Sci.*, 1980, **49**, 569.
33. H. Miyagawa, N. Fujikowa, H. Kohda, K. Yamasaki, K. Taniguchi and R. Tanaka, *Planta Med.*, 1986, **4**, 321.
34. M. V. Chalapathi, S. Thimmegowda, V. R. Rama Krishna Prama and T. G. Prasad, *Crop Res.*, 1997, **14**, 347.

35. R. Raina, S. K. Bhandari, R. Chand and Y. Sharma, *J. Med. Plants Res.*, 2013, 7, 1793.
36. J. Goettemoeller and A. Ching, in *Perspectives on New Crops and New Uses*, ed. J. Janick, ASHS Press, Alexandria, 1999, pp. 510–511.
37. M. Macchia, L. Andolfi, L. Ceccarini and L. G. Angelini, *Ital. J. Agron.*, 1997, **2**, 55.
38. A. Martini, S. Tavarini, M. Macchia, G. Benelli, A. Canale, D. Romano and L. G. Angelini, *Plant Biosyst.*, 2017, **151**, 341–351.
39. E. J. Kennelly, in *Stevia, the Genus of Stevia, Medicinal and Aromatic Plants-industrial Profiles*, ed. A. D. E. Kinghorn, Taylor and Francis, London, 2002, pp. 68–85, ISBN 0-415-26830-3.
40. S. Singh and G. Rao, *Sugar Tech*, 2005, **71**, 17.
41. J. E. Brandle and P. G. Telmer, *Phytochemistry*, 2007, **68**, 1855.
42. U. Wölwer-Rieck, B. May, C. Lankes and M. Wüst, *J. Agric. Food Chem.*, 2014, **62**, 2428.
43. J. E. Brandle and N. Rosa, *Can. J. Plant Sci.*, 1992, **72**, 1263.
44. S. Tavarini and L. G. Angelini, *J. Sci. Food Agric.*, 2013, **93**, 2121.
45. Y. S. Huang, A. G. Guo, Y. Qian, L. Y. Chen and H. F. Gu, *J. Plant Resour. Environ.*, 1995, **4**, 28.
46. C. Gardana, M. Scaglianti and P. Simonetti, *J. Chromatogr. A*, 2010, **1217**, 1463.
47. V. S. P. Chaturvedula, J. F. Clos, J. Rhea, D. Milanowski, U. Mocek, G. E. DuBois and I. Prakash, *Phytochem. Lett.*, 2011, **4**, 209.
48. U. Wölwer-Rieck, *J. Agric. Food Chem.*, 2012, **60**, 886.
49. J. E. Brandle, *Can. J. Plant Sci.*, 1999, **79**, 85.
50. M. R. Jenner, in *Progress in Sweeteners*, ed. T. Grenby, Elsevier, London, 1989, pp. 121–141.
51. L. Andolfi, M. Macchia and L. Ceccarini, *Ital. J. Agron.*, 2006, **1**, 257.
52. J. I. Lee, K. H. Kang, H. W. Park and Y. S. Ham, *Res. Rep. ORD*, 1982, **24**, 186.
53. C. Barbet-Massin, S. Giuliano, L. Alletto, J. Daydé and M. Berger, *Genet. Resour. Crop Evol.*, 2015, **63**, 685–694.
54. T. Morita, Dried leaves, *Jp. Pat.* 62–96025, 1987[English abstract].
55. S. Shizhen, *Sci. Agric. Sin.*, 1995, **28**, 37.
56. T. Morita and B. Yucheng, Variety of *Stevia rebaudiana* Bertoni, *US Pat.* 6031157, 1998.
57. I. Prakash, A. Markosyan and C. Bunders, *Foods*, 2014, **3**, 162.
58. http://www.foodnavigator-usa.com/Trends/Natural-sweeteners, accessed 30 January, 2018.
59. T. Lee, L. A. Nattress and W. Riha, *US Pat.* 2013/0071521 A1, 2013.
60. J. M. C. Geuns, *Phytochemistry*, 2003, **64**, 913.
61. S. Ghanta, A. Banerjee, A. Poddar and S. Chattopadhyay, *J. Agric. Food Chem.*, 2007, **55**, 10962.
62. V. Chatsudthipong and C. Muanprasat, *Pharmacol. Ther.*, 2009, **121**, 41.
63. H. Karaköse, R. Jaiswal and N. Kuhnert, *J. Agric. Food Chem.*, 2011, **59**, 10143.

64. R. Lemus-Mondaca, A. Vega-Galvez, L. Zura-Bravo and K. Ah-Hen, *Food Chem.*, 2012, **132**, 1121.
65. S. Tavarini, C. Sgherri, A. M. Ranieri and L. G. Angelini, *J. Agric. Food Chem.*, 2015, **63**, 7041.
66. C. Bender, S. Graziano and B. F. Zimmermann, *Int. J. Food Sci. Nutr.*, 2015, **66**, 553.
67. P. D. Caligari, *eLS*, 2001, DOI: 10.1038/npg.els.0002024.
68. R. E. Lorenzana and R. Bernardo, *Theor. Appl. Genet.*, 2009, **120**(1), 151.
69. S. Ceunen and J. M. C. Geuns, *Phytochemistry*, 2013, **89**, 32.
70. J. Metivier and A. M. Viana, *J. Exp. Bot.*, 1979, **30**, 1211.
71. E. I. Ermakov and A. A. Kochetov, *Russ. Agric. Sci.*, 1994, **11**, 11.
72. E. I. Ermakov and A. A. Kochetov, *Russ. Agric. Sci.*, 1996, **1**, 9.
73. H. Mizukami, K. Shiba, I. Satoshi and H. Ohashi, *Shoyakugaku Zasshi*, 1983, **37**, 175.
74. U. Tariq, M. Ali and B. H. Abbasi, *J. Photochem. Photobiol., B*, 2014, **130**, 264.
75. T. Lester, *Aust. New Crop Newsl.*, 1999, **11**, 1.
76. S. Ceunen, S. Werbrouck and J. M. C. Geuns, *J. Plant Physiol.*, 2012, **169**, 749.
77. P. K. Pal, R. Kumar, V. Guleria, M. Mahajan, P. Prasad, V. Pathania, B. S. Gill, D. Singh, G. Chand, B. Singh, R. D. Singh and P. S. Ahuja, *BMC Plant Biol.*, 2015, **15**, 67.
78. R. M. Moraes, M. A. Donega, C. L. Catrell, S. Mello and J. D. McChesney, *Ind. Crops Prod.*, 2013, **51**, 385.
79. P. K. Pal, M. Mahajan, R. Prasad, V. Pathania, B. Singh and P. S. Ahujac, *Ind. Crops Prod.*, 2015, **65**, 556.
80. L. G. Angelini and S. Tavarini, *Commun. Soil Sci. Plant Anal.*, 2014, **45**, 2577.
81. M. Karimi, A. Ahmadi, J. Hashemi, A. Abbasi, S. Tavarini, L. Guglielminetti and L. G. Angelini, *Sci. Hortic.*, 2015, **183**, 93.
82. J. W. P. Carneiro, in *Stevia rebaudiana (Bert.) Bertoni: produção de sementes*, U.E.M, Maringá, 1990, p. 65.
83. R. Kumar, S. Sharma, K. Ramesh, R. Prasad, V. L. Pathania, B. Singh and R. D. Singh, *Indian J. Agron.*, 2012, **57**, 74.
84. N. Taleie, Y. Hamidoghli, B. Rabiei and S. Hamidoghli, *Int. J. Agric. Crop Sci.*, 2012, **4**, 298.
85. I. Ibrahim, M. Nasr, B. Mohammed and M. El-Zefzafi, *Sugar Tech*, 2008, **10**, 248.
86. R. Monteiro, *Rev. Bras. Bot.*, 1982, **5**, 5[English abstract].
87. P. M. Magalhães, in *Fundamentos de Agrotecnología de Cultivo de Plantas Medicinales Iberoamericanas*, CYTED-CAB, Bogotá, 2000, pp. 441–450.
88. H. Nurhaimi and M. N. Toruan, *War. Pus. Penelit. Bioteknol. Perkeb.*, 1995, **1**, 2[English summary].
89. V. de Oliveira, E. R. Forni-Martins, P. M. Magalhães and M. N. Alves, *Genet. Mol. Biol.*, 2004, **27**(2), 215.
90. W. Handro, C. M. Ferreira and E. I. S. Floh, *Plant Sci.*, 1993, **93**, 1.

91. D. J. Midmore and A. H. Rank, in *A New Rural Industry – Stevia – To Replace Imported Chemical Sweeteners*, A report for the Rural Industries Research and Development Corporation, RIRDC Publication No W02/022, 2002.
92. S. Nakamura and Y. Tamura, *Jpn. J. Trop. Agric.*, 1985, **29**, 109–116.
93. R. Kasai, N. Kaneda, O. Tanaka, K. Yamasaki, I. Sakamoto, K. Morimoto, S. Okada, S. Kitahata and H. Furukawa, *Nippon Kagaku Kaishi*, 1981, **5**, 726(English abstract).
94. O. Tanaka, *Pure Appl. Chem.*, 1997, **69**, 675.
95. C. Lankes, U. M. Zabala and V. Müller, in *XXVIII International Horticultural Congress on Science and Horticulture for People (IHC2010)*, 2010, vol. 936, p. 273.
96. E. T. Anami, J. P. Poletine, M. C. Gonçalves-Vidigal, P. S. Vidigal Filho, G. F. Lacanallo, M. V. Kvitschal and A. Gonela, *J. Food Agric. Environ.*, 2010, **8**, 463.
97. M. W. Chalapathi, S. Thimmegowda and S. Sridhara, *Ind. Agric.*, 1998, **42**, 137.
98. M. W. Chalapathi, S. Thimmegowda, N. D. Kumar, G. G. E. Rao and J. Chandraprakash, *Crop Res.*, 1999, **21**, 947.
99. Y. T. Shyu, *J. Agric. Res. China*, 1991, **10**, 29.
100. F. Tateo, M. Mariotti, M. Bononi, E. Lubian, S. Martello and L. Cornara, *Ital. J. Food Sci.*, 1998, **10**, 261.
101. X. Y. Weng, J. Y. Sun and R. C. Zang, *J. Zhejiang Agric. Univ.*, 1996, **22**(5), 538.
102. A. P. Frederico, P. M. Ruas, M. A. Marin-Morales, C. F. Fuas and J. N. Nakajima, *Braz. J. Genet.*, 1996, **19**, 605.
103. R. Monteiro, Taxonomia e biologia da reproducao da *Stevia rebaudiana* Bert, PhD thesis, Univ. Estadual de Campinas, Brazil, 1980.
104. C. Darlington and A. P. Wylie, in *Chromosome Atlas of Flowering Plants*, G. Allen and Unwin Ltd., London, 2nd edn, 1955, p. 519.
105. Z. Bolkhoviskikh, V. Grif, T. Matvejeva and H. V. Zakharyeva, *Chromosome Number of Flowering Plants*, VL Komarov Botanical Institute, Academy of Sciences of the USSR, Moscou, 1969, p. 926.
106. R. J. Moore, Index to plant chromosome numbers 1967–1971, *Regnum Veg.*, 1973, **90**, 1.
107. R. J. Moore, Index to plant chromosome numbers 1972, *Regnum Veg.*, 1974, **91**, 1.
108. R. J. Moore, Index to plant chromosome numbers 1973–1974, *Regnum Veg.*, 1977, **96**, 1.
109. P. Goldblatt, in *Index to Plant Chromosome Numbers 1975–1978*, Monographs in Systematic Botany from the Missouri Botanical Garden, Missouri Botanical Garden, St. Louis, 1981, vol. 5, p. 553.
110. P. Goldblatt, in *Index to Plant Chromosome Numbers 1979–1981*, Monographs in Systematic Botany from the Missouri Botanical Garden, Missouri Botanical Garden, St. Louis, 1984, vol. 8, p. 432.
111. P. Goldblatt, in *Index to Plant Chromosome Numbers 1982–1983*, Monographs in Systematic Botany from the Missouri Botanical Garden, Missouri Botanical Garden, St. Louis, 1985, vol. 13, p. 224.

112. P. Goldblatt, in *Index to Plant Chromosome Numbers 1984–1985*, Monographs in Systematic Botany from the Missouri Botanical Garden, Missouri Botanical Garden, St. Louis, 1988, vol. 23, p. 264.
113. P. Goldblatt and D. E. Johnson, in *Index to Plant Chromosome Numbers 1986–1987*, Monographs in Systematic Botany from the Missouri Botanical Garden, Missouri Botanical Garden, St. Louis, 1990, vol. 30, p. 242.
114. P. Goldblatt and D. E. Johnson, in *Index to Plant Chromosome Numbers 1988–1989*, Monographs in Systematic Botany from the Missouri Botanical Garden, Missouri Botanical Garden, St. Louis, 1991, vol. 40, p. 238.
115. P. Goldblatt and D. E. Johnson, in *Index to Plant Chromosome Numbers 1992–1993*, Monographs in Systematic Botany from the Missouri Botanical Garden, Missouri Botanical Garden, St. Louis, 1996, vol. 58, p. 276.
116. P. Goldblatt and D. E. Johnson, in *Index to Plant Chromosome Numbers 1994–1995*, Monographs in Systematic Botany from the Missouri Botanical Garden, Missouri Botanical Garden, St. Louis, 1998, vol. 69, p. 208.
117. J. L. Grashoff, M. W. Burner and D. K. Northington, *Brittonia*, 1972, **25**, 379.
118. D. D. Soejarto, in *Stevia – The Genus Stevia (Medicinal and Aromatic Plants – Industrial Profiles)*, ed. A. D. Kinghorn, Taylor & Francis/CRC Press, New York/London, UK, 2002, pp. 40–67.
119. R. M. King and H. Robinson, *The Genera of the Eupatorieae (Asteraceae)*, Monographs in Systematic Botany, Missouri Botanical Garden, St. Louis, MO, 1987.
120. R. K. Jansen, T. F. Stuessy, S. Dias-Piedrahita and V. A. Funk, *Caldasia*, 1984, **14**, 7.
121. J. R. Coleman, *Rhodora*, 1968, **70**, 228.
122. N. G. Galiano, *Darwiniana*, 1987, **28**, 311.
123. N. Galiano and J. H. Hunziker, *Darwiniana*, 1987, **28**, 1.
124. L. G. Angelini, G. Leone, C. Russo and S. Tavarini, in *Knowledge on Tour in Europe*, ed. J. M. C. Geuns, Proceedings of the 7th Stevia Symposium of European Stevia Association (EUSTAS), 24–26 June 2013, Toulouse, pp. 93–96, ISBN: 9789074253277.
125. C. Hastoy, P. Cosson, E. Sclaunich, P. Jannot, P. Boutié, D. Rolin and V. Schurdi-Levraud, in *Stevia: Growth in Knowledge and Taste*, ed. S. Ceunen and J. M. C. Geuns, Proceedings of the 8th Stevia Symposium of European Stevia Association (EUSTAS), 27–29 Jan 2015, p. 225, ISBN: 9789074253291 2015.
126. A. K. Yadav, S. Singh, B. Singh, R. K. Sharma, S. Vijaylata, A. Sood and P. S. Ahuja, Release of *Stevia rebaudiana* cultivar HIM Stevia (CSIR-IHBT-ST-01), 2014, http://www.ihbt.res.in.
127. P. D. Caligari and B. P. Forster, *eLS*, 2015, 1–11.
128. J. Brown, P. D. S. Caligari and H. A. Campos, in *Plant Breeding*, ed. J. Brown, P. Caligari and H. Campos, Wiley-Blackwell, 2nd edn, 2014.

129. S. Sharma, S. Waila, S. Singh and R. Kumar, *J. Sci. Food Agric.*, 2016, **96**, 1867.
130. J. I. Lee, K. H. Kang and Y. S. Ham, *Annu. Res. Rep.*, 1978, **21**, 21.
131. S. Z. Shu, *Sci. Agric. Sin.*, 1995, **28**, 37.
132. L. Meienberg, T. Sommer, M. Lebrecht, S. Lovera, B. Gonzalez, V. Luig, K. Bremen, M. Steiner, M. Glauser and U. Kienle, in *The Bitter Sweet Taste of Stevia*, Berne Declaration, CEIRAD, Misereor, Pro Stevia Switzerland, SUNU, University of Hohenheim, 2015, http://bernedeclaration.ch/fileadmin/files/documents/Biodiversitaet/FINAL_EN_STEVIA_REPORT_20151112_letzte_Korr.pdf, Accessed 15 Jan 2016.
133. E. A. Sys, A. A. Marsolais and J. Brandle, Stevia plant named 'RSIT 94-1306', *US Pat.* PP10562, 1998.
134. A. A. Marsolais, J. Brandle and E. A. Sys, Stevia plant named 'RSIT 94-751', *US Pat.* PP10564, 1998.
135. A. Luwańska, L. Perz, G. Mańkowska and K. Wielguset, *Kerva Pol.*, 2015, **61**, 50.
136. A. R. Hallauer and L. L. Darrah, *Crit. Rev. Plant Sci.*, 1985, **3**, 1.
137. J. E. Brandle, *US Pat.* 6255557 B1, 2001.
138. N. T. P. Thao, K. Ureshino, I. Miyajima, Y. Ozaki and H. Okubo, *Plant Cell, Tissue Organ Cult.*, 2003, **72**, 19–25.
139. C. N. Rameshsing, S. N. Hegde, M. R. Wallalwar and M. Vasundhara, *Res. Environ. Life Sci.*, 2015, **8**, 393.
140. A. K. Yadav, S. Singh, S. C. Yadav, D. Dhyani, G. Bhardwaj, A. Sharma and B. Singh, *Indian J. Agric. Sci.*, 2013, **83**, 159.
141. S. Kokusaku, New triploid of *Stevia rebaudiana* Bertoni contains sweet diterpenoid, *Jp. Pat.* 2242622-A; 2748141-B2, 1990.
142. H. Shuichi, Y. Tsuneo and F. Satoshi, *Jpn. J. Trop. Agric.*, 2001, **45**, 281.
143. C. Shaopan, C. Shaoqiu, Y. Yinghua and Zhaizhiqiang, *J. Wuhan Bot. Res.*, 1995, **13**(1), 1.
144. S. N. Hedge, C. N. Rameshsing and M. Vasundhara, *Med. Plants – Int. J. Phytomed. Rel. Ind.*, 2015, **7**, 188.
145. E. Flachsland, L. Mroginski and J. Davina, *Biocell*, 1996, **20**, 87.
146. G. V. Garnighian, Stevia plant named 'T60', *US Pat.* PP22,593 P3, 2012.
147. E. Francia, G. Tacconi, C. Crosatti, D. Barabaschi, D. Bulgarelli, E. Dall'Aglio and G. Valè, *Plant Cell, Tissue Organ Cult.*, 2005, **82**, 317.
148. I. Schuster, *Crop Breed. Appl. Biotechnol.*, 2011, **S1**, 50.
149. M. Thiyagarajan and P. Venkatachalam, *Ann. Phytomed.*, 2015, **4**(1), 79.
150. Y. Yao, M. Ban and J. Brandle, *Genome*, 1999, **42**, 657.
151. K. Chester, E. T. Tamboli, R. Parveen and S. Ahmad, *Pharm. Biol.*, 2013, **51**, 771.
152. T. Nakai, T. Konishi, X. Q. Zhang, R. Chollet, N. Tonouchi, T. Tsuchida, F. Yoshinga, H. Mori, F. Sakai and T. Hayashi, *Plant Cell Physiol.*, 1998, **39**, 1337.
153. N. Toruan-Mathius, T. Pratiwi and T. Hutabarat, *Menara Perkebunan*, 1995, **63**, 33.

154. S. A. Khan, L. ur Rahman, R. Verma and K. Shanker, *Acta Physiol. Plant.*, 2016, **38**(1), 1.
155. J. E. Brandle, A. Richman, A. K. Swanson and B. P. Chapman, *Plant Mol. Biol.*, 2002, **50**, 613.
156. A. S. Richman, M. Gijzen, A. N. Starratt, Z. Yang and J. E. Brandle, *Plant J.*, 1999, **19**, 411.
157. A. Richman, A. Swanson, T. Humphrey, R. Chapman, B. McGarvey, R. Pocs and J. Brandle, *Plant J.*, 2005, **41**, 56.
158. H. P. Singh, S. K. Dhir and S. Dhir, in *Stevia – Compendium of Transgenic Crop Plant*, 2008.
159. M. Mubarak, Y. El Halmouch, A. Belal, M. Abd Elfadeel, T. Nasr EL-Din, S. F. Mahmoud, M. E. El Sharnouby and E. El Sarag, *POJ*, 2015, **8**(3), 232.
160. C. J. Thompson, N. R. Movva, R. Tizard, R. Crameri, J. E. Davies, M. Lauwereys and J. Botterman, *EMBO J.*, 1987, **6**(9), 2519.
161. R. A. Jefferson, T. A. Kavanagh and M. W. Bevan, *EMBO J.*, 1987, **6**(13), 3901.
162. S. B. Gelvin, *Microbiol. Mol. Biol. Rev.*, 2003, **67**(1), 16.
163. E. W. Nester, M. P. Gordon, R. M. Amasino and M. F. Yanofsky, *Annu. Rev. Plant Physiol.*, 1984, **35**, 387.
164. A. N. Binns and M. F. Thomashow, *Annu. Rev. Microbiol.*, 1988, **42**, 575.
165. S. A. Khan, L. U. Rahman, K. Shanker and M. Singh, *Protoplasma*, 2014, **251**(3), 661.
166. P. Guleria and S. K. Yadav, *PLoS One*, 2013, **8**, e74731.
167. P. Waterhouse and C. A. Helliwell, *Nat. Rev.*, 2003, **4**, 29.
168. J. Zupan, T. R. Muth, O. Draper and P. Zambryski, *Plant J.*, 2000, **23**, 11.
169. S. K. Yadav and P. Guleria, *Crit. Rev. Food Sci.*, 2012, **52**, 988.
170. UPOV, PLUTO – plant variety database, http://www3.wipo.int/pluto/en/, Accessed 23 June 2015.

CHAPTER 2

Biosynthesis of Steviol Glycosides and Related Diterpenes in Leaves and Glandular Trichomes of Stevia rebaudiana *Bertoni*

A. OEHME[a], W. SCHWAB[b] AND M. WÜST*[a]

[a]Rheinische Friedrich-Wilhelms-Universität Bonn, Institute of Nutritional and Food Sciences, Bioanalytics, Endenicher Allee 11-13, Bonn, 53115, Germany; [b]Technische Universität München, Center of Life and Food Sciences Weihenstephan, Biotechnology of Natural Products, Liesel-Beckmann-Str. 1, Freising, 85354, Germany
*E-mail: matthias.wuest@uni-bonn.de

2.1 Introduction

Steviol glycosides are terpenoid compounds, composed of a tetracyclic diterpenoid (*ent*-kaurenoic) backbone, the so-called steviol, which forms O-glycosidic and ester bonds to different sugars.

The biosynthetic pathway of steviol glycosides has been elucidated in large parts up to now. Early steps of the synthesis involve the formation of the

Food Chemistry, Function and Analysis No. 7
Steviol Glycosides: Cultivation, Processing, Analysis and Applications in Food
Edited by Ursula Wölwer-Rieck

Published by the Royal Society of Chemistry, www.rsc.org

isoprenoid basic units *via* the deoxyxylulose phosphate (DXP) pathway,[3,4,12] a mevalonic acid independent pathway, which has been located in the plastids.[5,6]

It further shares four early steps of the biosynthesis of gibberellins (GA),[7] which are known to be phytohormones with endogenous growth regulatory function.[8–10] The function of steviol glycosides themselves in the plant, however, is not yet clarified, although many suggestions had been made until now, including the gibberellin like or inhibiting activity of the aglycone steviol,[11–13] feeding deterrent activity,[14,15] function as osmolyte[12] or oxidative stress inhibitory activity.[16,17]

In this chapter, we will give an overview of the synthesis of steviol glycosides, with respect to the metabolic cross-talk of the two distinct pathways, responsible for the formation of the isoprene basic C_5-building blocks and we will go into detail concerning the UDP-glycosyltransferases, enzymes that catalyse the formation of numerous steviol glycosides.

Furthermore, we will give details about the morphology of the *Stevia rebaudiana* Bertoni plant and particularly discuss the plant's hairiness. Those hairs, also called trichomes, do not only cover both, upper (*adaxial*) and bottom side (*abaxial*) of the leaf, but all vegetative and reproductive parts of the plant, such as corolla, stem and achene.[18]

Plant trichomes drew very much attention in the last years, due to their capability of producing and storing high amounts of secondary plant compounds such as terpenes, which are the most abundant and diverse class of plant metabolites.[19–23]

These findings raise the question whether steviol glycosides are produced and stored in the plant's trichomes as well. Finding steviol glycosides in trichomes could be of interest for breeding purposes and may also improve the extraction and manufacturing of the natural sweetener by reducing the multistep purification process.[24] This processing is necessary, as only the use of highly purified steviol glycosides is legalised and regulated in many countries at this time.[25,26] However, along with this complex manufacturing process, which includes protein precipitation and treatments with activated carbon, ion-exchange resin and organic solvents[24] to remove other plant constituents, the natural authenticity of extracted steviol glycosides is questioned. Thus, extraction and purification of steviol glycosides from trichome cavities might be less elaborate.

Indeed, there is some evidence that steviol glycosides may accumulate in the glandular trichomes of the leaves, as previous investigations have shown that the content of steviol glycosides positively correlates to trichome density. It has been reported that steviol glycoside content is at a maximum in younger leaves.[27] According to transmission electronic microscopy investigation, trichome density is highest in young and developing leaves, while their number decreases during senescence.[28]

Recent comparative studies of isolated trichomes and the remainder leaves using transcriptomic methods and liquid chromatography UV-analyses (LC-UV), however, gave evidence that steviol glycoside storage and production is limited to mesophyll cells. Although genes encoding for enzymes, which are involved in steviol glycoside synthesis, were expressed in trichomes, no steviol glycosides could be detected in isolated trichomes *via* LC-UV analysis.[29] However, recent GC-MS and LC-MS/MS analyses proved the presence of the intermediate *ent*-kaurene and of steviol glycosides themselves in isolated foliar trichomes.[73,132]

2.2 Morphology of *Stevia rebaudiana* Bertoni

Stevia rebaudiana Bertoni is a perennial herb which belongs to the Asteraceae family (former Compositae). Characteristics for *S. rebaudiana* are filamentary roots and a thin stem, which is up to 80 cm tall in cultivation and produces secondary shoots from the base. Often, cultivated plants are more strongly branched than stevia plants in their wild habitat.

Leaves are growing directly from the stem (*subsessile*). They are simple, opposite, with alternating leaf nodes, up to 3 cm long and toothed on the upper part. Whereas young leaves have a lanceolate shape with serrated margins and an acute apex, older leaves become more cuneate and wavy-toothed (*crenate*) with obtuse apices (see Figure 2.1).

There are three primary veins arising from the leaf base, which are submerged on the upper (*adaxial*) and bulging from the lower (*abaxial*) side, with reticulate secondary veinlets. The green colour of the adaxial leaf side is more intensive compared to the green of the abaxial side.

Figure 2.1 Image of the shoot apex illustrates the subsessile, opposite and alternating leaves with lanceolate shape, serrated margins and an acute apex.

The capitulum is arranged as lax paniculate-corymbous inflorescences at the ends of branches, with lanceolate involucral bracts covering five disk florets. The florets appear as slender white corolla tubes.[30]

The fruits of *Stevia rebaudiana* are spindle shaped achenes with persistent pappi, so called cypselas, with seeds of 3.5–4 mm in length.[31]

2.2.1 Foliar Trichomes of *Stevia rebaudiana* Bertoni

Trichomes are fine outgrowths of epidermal cells on the surfaces of plants, differing in shape and function. There are two main types of trichomes to be distinguished: 'glandular' and 'non-glandular' trichomes. Whereas, in general, non-glandular trichomes mainly vary in their morphological structure; glandular trichomes, also called oil glands,[32] characteristically differ in the substances they are accumulating and/or secreting.[33] Hence, glandular trichomes are anatomical structures, specialised in the biosynthesis of their secretions.[34]

Both glandular and non-glandular trichomes may contribute to plant protection mechanisms, comprising protection against herbivores, pathogens, extreme light and temperature. While non-glandular trichomes more likely serve as a physical barrier, glandular trichomes may also deter by releasing e.g. herbicidal, antimicrobial and insecticidal secretions, which they are accumulating.[33,35–39]

Vegetative and reproductive organs of *Stevia rebaudiana* Bertoni, such as stems (see Figure 2.2), flower petals, sepals, stamen (see Figure 2.3), leaves (see Figure 2.4) and achenes, are covered with non-glandular and glandular trichomes.[18]

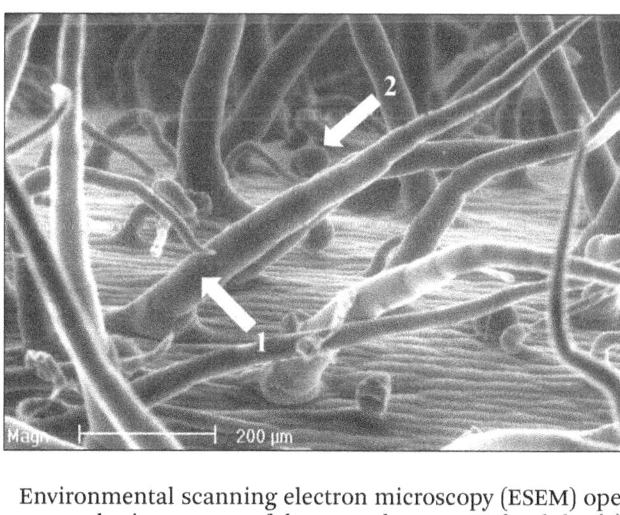

Figure 2.2 Environmental scanning electron microscopy (ESEM) operating under atmospheric pressure of the stem shows non-glandular (1) and glandular (2) trichomes.

Figure 2.3 ESEM-pictures of *Stevia rebaudiana*, genotype Gawi, of the capitulum (A) and close-ups of involucral bracts (B), corolla (C) and stamen (D) show holistic hairiness.

Two kinds of non-glandular trichomes, large and short ones, and one kind of a glandular trichome cover adaxial and abaxial leaf sides of stevia (see Figures 2.4 and 2.5).[18]

Whereas the large non-glandular foliar trichomes are up to 500 μm long, composed of 7–16 cells arranged in a row (*uniseriate*) and have a pointing tip, the shorter non-glandular foliar trichomes consist of only 5–8 uniseriate cells, with a length of up to 150 μm, ending with a rounded tip.[28]

Glandular foliar trichomes are sunken into the epidermal surface. They are built of ten cells, which are, in contrast to the non-glandular hairs, arranged in pairs (*biseriate*): four cells forming the two subapical pairs, namely basal cells and stalk cells, and six secretory cells, forming three apical pairs.[18,28,40] Plasmodesmata, channels that connect neighbouring cells, have been found to connect these pair cells with mesophyll cells.[41]

The tip of the glandular trichomes is bulged and builds a shield with a diameter of up to 60 μm, as a result of the secretions which accumulate underneath the cuticle (*trichome cavity* or *subcuticular cavity*).[18,28]

By the break of the cuticle sheath at a weak line, visible in Figure 2.6, secretions underneath the cuticle are released.[18]

Involucral bracts show a comparable distribution of uniseriate long and short non-glandular trichomes and glandular trichomes. Glandular trichomes are not sunken into the epidermal surface. The corolla shows a slight variation in the trichome pattern: whereas dorsal sides show long uniseriate non-glandular and biseriate non-sunken glandular trichomes,

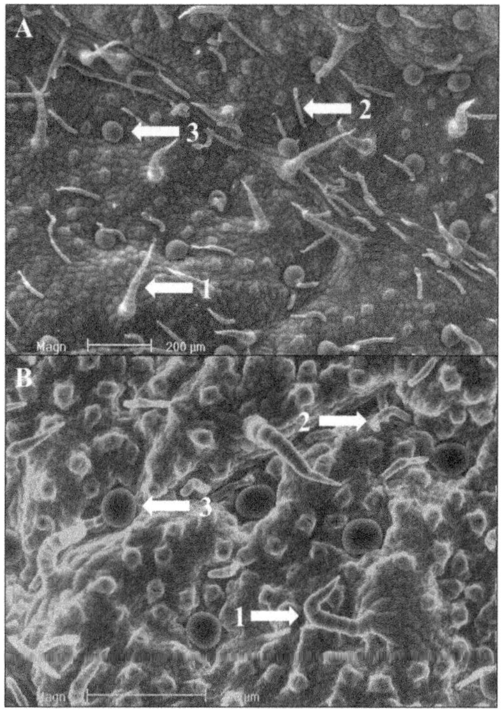

Figure 2.4 Electron microscopy under atmospheric pressure of the (A) adaxial and (B) abaxial leaf surface of *Stevia rebaudiana*, genotype Gawi, shows two types of non-glandular trichomes (long (1) and short (2) hairs) and one type of a glandular trichome, which is sunken into the epidermal surface (3).

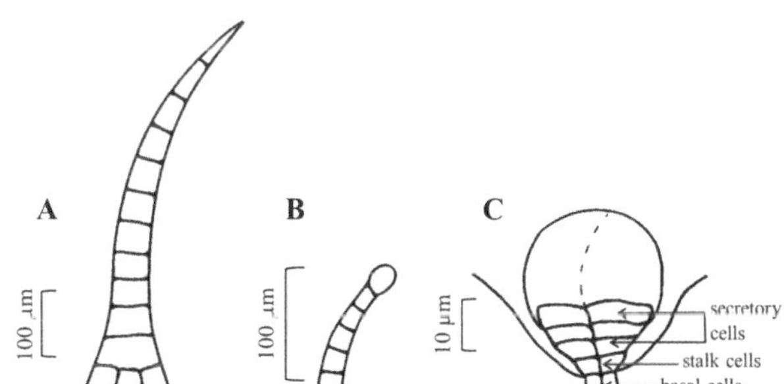

Figure 2.5 Schematic diagram of non-glandular (A) and (B), and glandular trichomes (C) of abaxial and adaxial sides of the stevia leaf. Inspired by Tateo *et al.*[18]

only long non-glandular hairs were found on the ventral side of the corolla throat. Former investigations prove that achenes even have stalked glandular trichomes next to twin hairs which are separated at the tip only (not pictured here).[18] Further electron microscopy reveals that the hairiness of the stem is

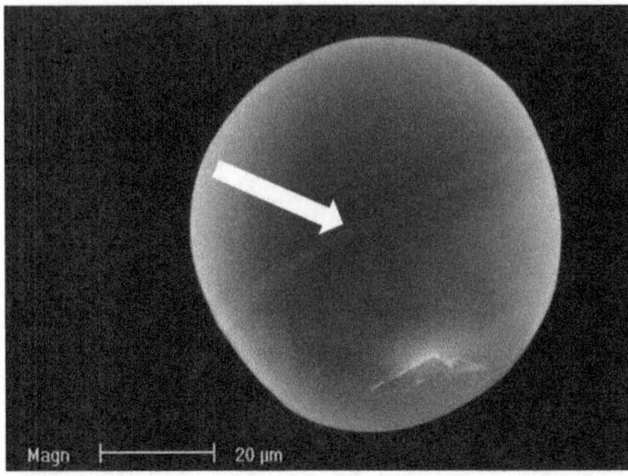

Figure 2.6 Electron microscopy of an isolated foliar glandular trichome. The arrow marks the weak line, where the cuticle sheath breaks to release accumulated secretions.

composed of long and short uniseriate non-glandular and stalked glandular trichomes as well (see Figure 2.2).

Due to the finding that leaves accumulate the highest content of steviol glycosides when compared to other organs like roots, flower, seeds and stem,[27,42] the leaves are the most intensively researched with regard to their glandular trichomes.[4,18,28,29,40,41,43] Ultrastructural observations revealed that the cells of foliar glandular trichomes are completely metabolically equipped, containing vacuoles, large nuclei, ribosomes, mitochondria, plastids, endoplasmic reticulum and Golgi apparatus elements.[28,40,41] However, those cell pairs are on different ontogenetic stages. Whereas basal cells have typical ultrastructure and active metabolism, reflected by containing a large nucleus, plastid apparatus and mitochondria and a dense cytoplasm with small vacuoles and electron dense inclusions, an increasing destruction is observed the more apical the trichome cells are situated.[28]

Besides, younger upper stevia leaves bear more glandular trichomes compared to bottom senile leaves.[28] This comes along with the fact that younger leaves accumulate higher contents of steviol glycosides, with the highest amount detected shortly before flowering.[27,42]

However, many studies have proven not only a great capacity of accumulating and storing secretions but also metabolic activity of glandular trichomes of sundry plant families. *In situ* biosynthesis of the monoterpene (−)-carvone in glandular trichomes of spearmint (*Mentha spicata*),[19] of phenylpropenes in peltate glandular trichomes of basil (*Ocimum basilicum*),[44] of acyl sugars in glandular trichomes of cultivated tomato (*Solanum lycopersicum*)[45] and of the labdane diterpene *Z*-abienol in tobacco (*Nicotiana tabacum*) glandular trichomes[46] was shown.

Taking into account, that steviol glycoside content and trichome density positively correlates,[28,42] it is a reasonable question whether steviol glycosides are produced and stored in the trichome's cavity as well. Analytical studies of *Stevia rebaudiana* glandular trichomes, however, have revealed that steviol glycoside production seems to be limited to mesophyll cells, as steviol glycosides were not detectable in isolated trichomes *via* LC-UV. In the same work, however, transcriptomic analyses proved that genes encoding enzymes of the methylerythritol pathway and steviol glycosides synthesis were expressed in glandular trichomes, but there were differences in the relative expression levels compared to mesophyll cells.[29]

2.2.1.1 Isolation Techniques for Trichomes

As first experiments with excised epidermis of *Majorana hortensis* showed the incorporation of isotopically labelled carbons of [U–^{14}C]-sucrose into monoterpenes, it was stated in 1977 that monoterpene biosynthesis is primarily located in epidermal tissue, in particular in epidermal oil glands.[47] Further metabolite analyses and modern transcriptomic studies followed proving the presence and biosynthesis of (−)-menthol, menthone and many other monoterpenes in peltate trichomes of peppermint leaves (*Mentha piperita*)[19,48] and the biosynthesis of flavonoids in alfalfa (*Medicago sativa* L.) glandular trichomes.[49] In addition, secondary metabolism and resistance-related gene expression in trichomes of *Nicotiana tabacum*,[50] upregulation of genes expressing the production of artemisinin, an anti-malarial drug, in glandular trichomes of *Artemisia annua*[51,52] and accumulation of diverse acyl sugar metabolites and monoterpene synthesis in trichomes of tomato leaves (*Solanum lycopersicum*)[21,45,53] was shown.

Discoveries like these required an optimisation in trichome isolation, as analysis of trichomes often is limited due to the difficulty of their isolation. Thus, over time, several mechanised techniques were developed to isolate epidermal structures, intact trichomes and their exudates. One of the most convenient is the extraction of exudates on the leaf surface by shortly dipping the leaves into an appropriate solvent, *e.g.* hexane, methanol or methyl-*tert*-butyl ether.[21,45,54–56] Other techniques involve abrasion of the leaf surface by scraping, cutting or brushing the trichomes off the leaves with *e.g.* a cover glass, a thin wire or a soft toothbrush.[57–62] Similar to these techniques the trichomes can be removed by pressing an adhesive tape onto the leaf surface and afterwards removing the trichomes off the tape by washing.[63] Just as common is the abrasion of trichomes with glass beads in a buffer medium.[44,64–66] Laser microdissection, as a targeted technique, has been recently used to cut trichomes off and to ensure their integrity for further transcriptomic studies.[67,68] Separation of glandular and non-glandular trichome mixtures is possible by sieving the isolated trichomes through fine mesh-nets or by gradient centrifugation.[59,65,69,70]

Depending on the analytical technique, it is further possible to collect trichome exudates by piercing trichome cavities with *e.g.* the tip of a solid phase micro extraction (SPME) fibre following by a GC-MS analysis.[71,72] Further, it has been proven that headspace (HS) GC-MS analyses of isolated trichomes

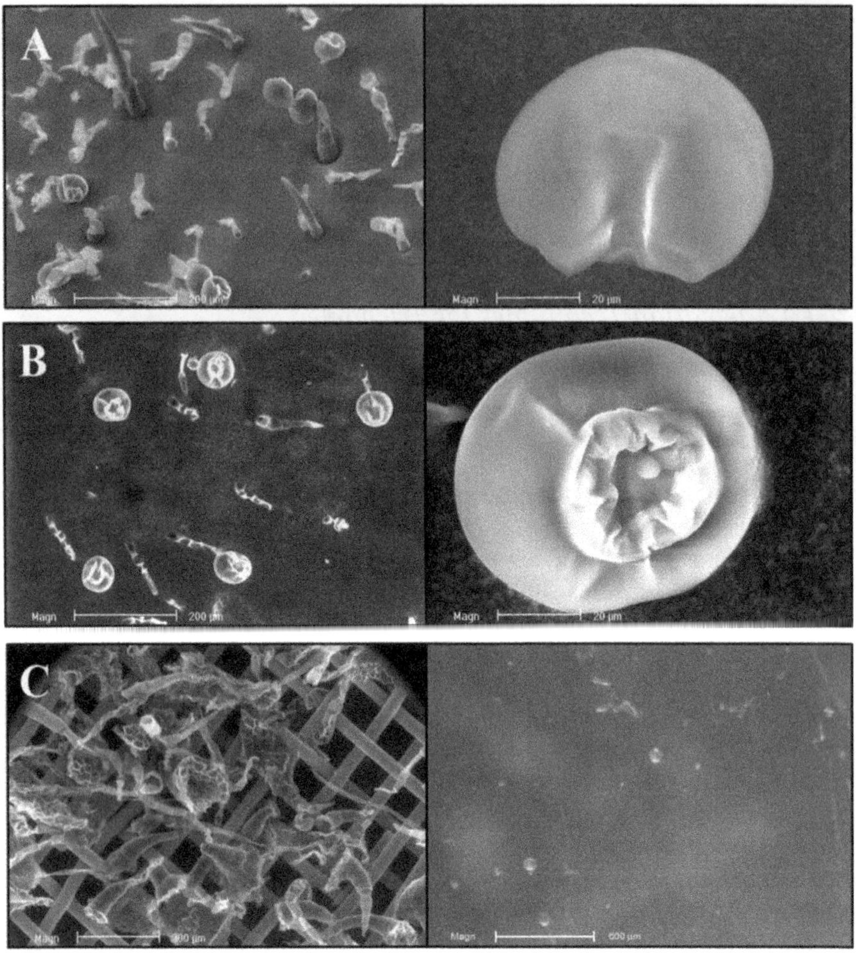

Figure 2.7 ESEM overview and detailed view of different techniques to isolate foliar trichomes of stevia leaves. Brushed off trichomes (A), isolated trichomes upon an adhesive tape which was pressed onto leaf surface (B), epidermal residue on mesh and isolated trichomes in filtrate solution (light microscope) (C).

and whole leaves are comparable, *i.e.* volatiles measured in headspaces of leaves most likely originate from trichomes.[4,22,29]

Isolating glandular trichomes of stevia leaves faces a particular difficulty, as the trichomes are sunken into epidermal indentations (see Figures 2.4 and 2.5). Thus, several techniques were tested and validated *via* electron microscopy at atmospheric pressure and *via* light microscopy.[73] The methods included: (A) brushing off the trichomes of adaxial and abaxial leaf sides with a painter's bristle brush after shortly freezing the leaves in liquid nitrogen; (B) pressing an adhesive tape onto the leaf surface and (C) using the mechanised glass beads technique, all following former publications (see Figure 2.7).[56,63,74] All methods, which were tested, achieved a satisfying separation of trichomes from the leaves. However, for analytical purposes, the most suitable technique was to remove the stevia trichomes by brushing them off the leaf surface, yielding a concentrated sample.

2.3 Biosynthesis of Steviol Glycosides

Steviol glycosides are diterpenes with an *ent*-kaurenoic backbone, called steviol, which forms O-glycosidic bonds *via* its hydroxyl group and ester bonds to carbohydrates *via* its carboxyl function. More than 35 steviol glycosides have been structurally identified, differing in the number and structure of attached sugar units and the sweet and bitter taste sensations.[1,12,75]

As all diterpenes, they arise from the C_5 units isopentenyl diphosphate (also isopentenyl pyrophosphate) (IPP) and dimethylallyl diphosphate (also dimethylallyl pyrophosphate) (DMAPP). The connection of one DMAPP to three IPP units *via* a three-step condensation is catalysed by the prenyl transferase enzyme geranylgeranyl diphosphate synthase (GGDPS) and forms geranylgeranyl diphosphate (GGPP).[7,76] The condensation starts with the ionisation of DMAPP by the loss of the pyrophosphate moiety and electrophilic addition of IPP to the allylic cation of DMAPP. Geranyl diphosphate (GPP) is generated which is again activated by the loss of the pyrophosphate moiety. Another IPP unit is electrophilically added and farnesyl diphosphate (FPP) is generated. By the addition of the last IPP unit to the activated farnesyl-cation GGPP is finally generated (see Figure 2.8).[77,78]

Terpene cyclases have been found to catalyse the two subsequent steps of steviol glycoside biosynthesis in stevia.[7,8] Protonation of a double bond of GGPP, initiated by copalyl diphosphate synthase (CDPS), induces cyclisation reactions, which are terminated by a loss of a proton at the methyl function, leading to copalyl diphosphate (CPP) (see Figure 2.9). Further on, kaurene synthase (KS) catalyses a sequence of cyclisation reactions: CPP is activated by the loss of diphosphate, inducing the cyclisation reactions to a three-ring and subsequently a four-ring system. A Wagner-Meerwein 1,2-alkyl shift followed by the loss of a proton eventually generates *ent*-kaurene, not only a precursor of steviol but also of gibberellic acid (see Figure 2.8).[8,77,79]

Kaurene oxidase (KO), a cytochrome P450 mono-oxygenase, which is located in the endoplasmic reticulum, is then catalysing hydroxylation and oxidation reactions, forming kaurenoic acid *via* *ent*-kaurenol and *ent*-kaurenal (see Figure 2.9).[80,81]

Requiring NADPH and molecular oxygen, another cytochrome P450 enzyme, namely kaurenoic acid 13-hydroxylase (KAH), catalyses the formation of steviol from kaurenoic acid (see Figure 2.9).[7,82,83]

Lastly, using uridine diphosphate (UDP) activated sugars, UDP-glycosyltransferases (UGTs) promote the glycosylation of the hydroxyl group at C-13 and carbon function at C-19 of steviol resulting in various steviol glycosides, differing in number and kind of sugars attached (see Figure 2.9).[13,84–86]

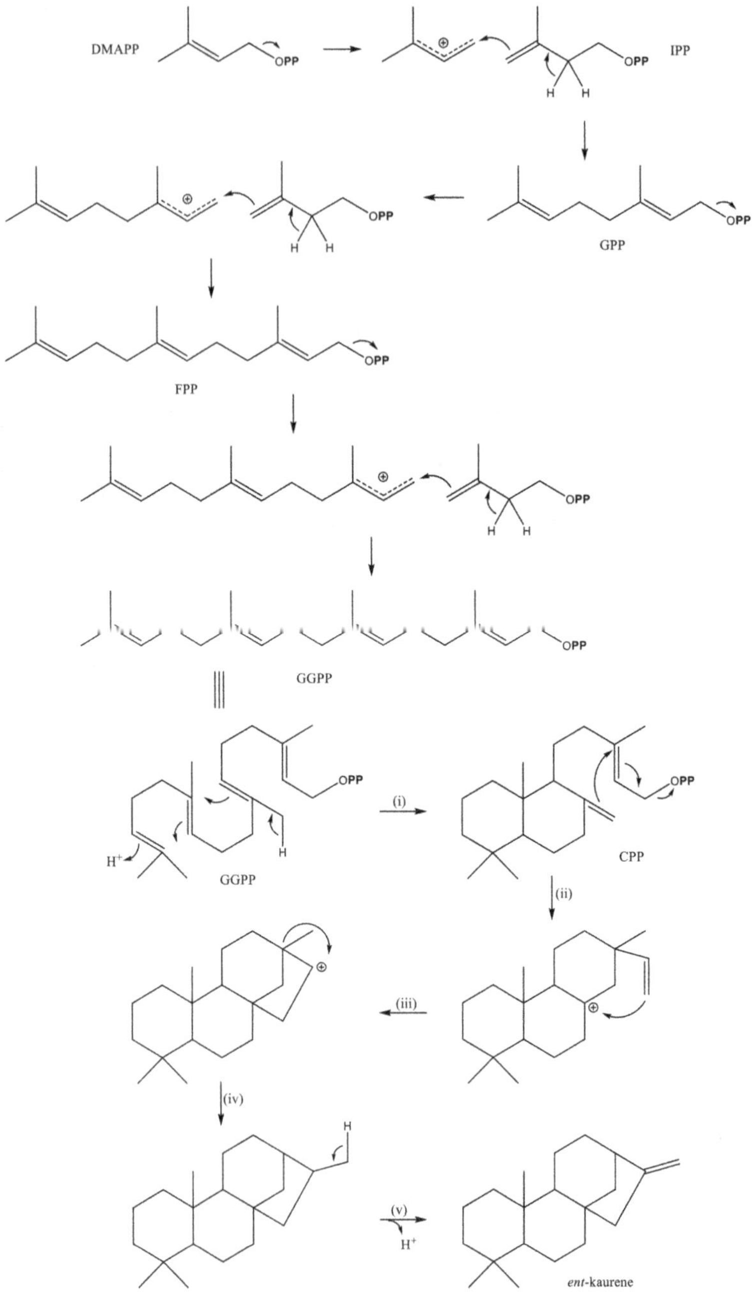

Figure 2.8 Formation of geranylgeranyl pyrophosphate (GGPP) *via* geranyl pyrophosphate (GPP) and farnesyl pyrophosphate (FPP), by the gradual condensation of the activated dimethylallyl pyrophosphate (DMAPP) cation to three isopentenyl pyrophosphate (IPP) units, catalysed by GGDP-synthase. The terpene cyclases (i) copalyl pyrophosphate synthase (CDPS) and (ii)–(v) kaurene synthase (KS) further catalyse cyclisation reactions of geranylgeranyl pyrophosphate (GGPP) to form copalyl diphosphate (CPP) and *ent*-kaurene.

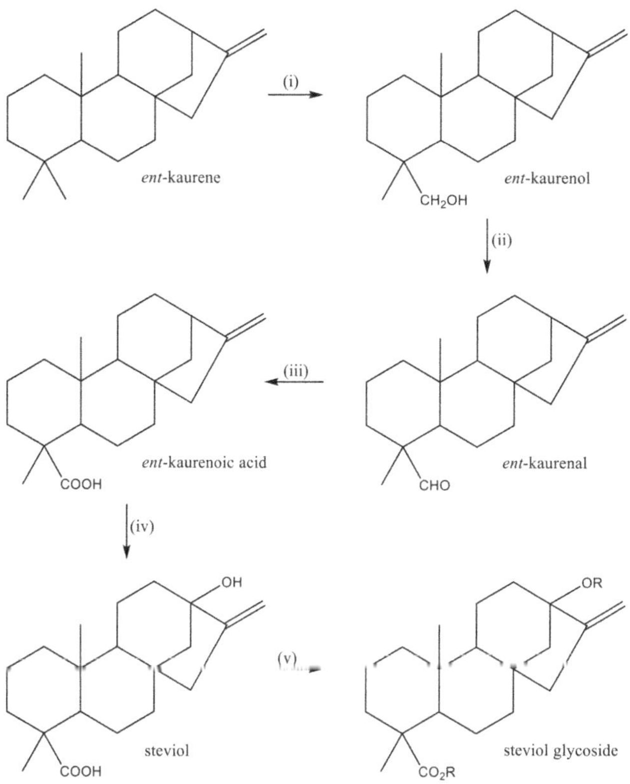

Figure 2.9 Cytochrome P450 mono-oxygenases (i)–(iii) kaurene oxidase (KO) and (iv) kaurenoic acid 13-hydroxylase (KAH) promote the oxidation of *ent*-kaurene to steviol *via* three intermediates (*ent*-kaurenol, *ent*-kaurenal, *ent*-kaurenoic acid). (v) Uridine diphosphate-glycosyltransferases (UTGs) eventually catalyse glycosylation reactions forming a variety of different steviol glycosides.

2.3.1 UDP-glycosyltransferases in *Stevia rebaudiana*

Comprehensive chemical analyses of the *Stevia rebaudiana* plant revealed the structure of more than 35 steviol glycosides.[134] Most of them elicit both sweet and bitter taste sensations. The glycone chain length, pyranose substitution and the C16 double bond are structural features determining the gustatory profile of steviol glycosides (see Figure 2.10).

Rebaudioside D and M carry five and six glucose units, respectively, and are especially appreciated because they have low sweetness thresholds and exhibit reduced bitter taste compared to the other stevioside derivatives. However, the most abundant steviol glycosides are 1,2-stevioside, rebaudioside A and C constituting 5–10%, 2–4% and 1–2% of the leaf dry weight, respectively. A combinatorial biosynthetic pathway involving at least four UGTs produces the mixture of differently glycosylated steviol glycosides. UGT85C2 and UGT74G1 transfer D-glucose units to the steviol backbone at the C_{13}-hydroxyl and C_{19}-carboxylic acid functional group, respectively giving

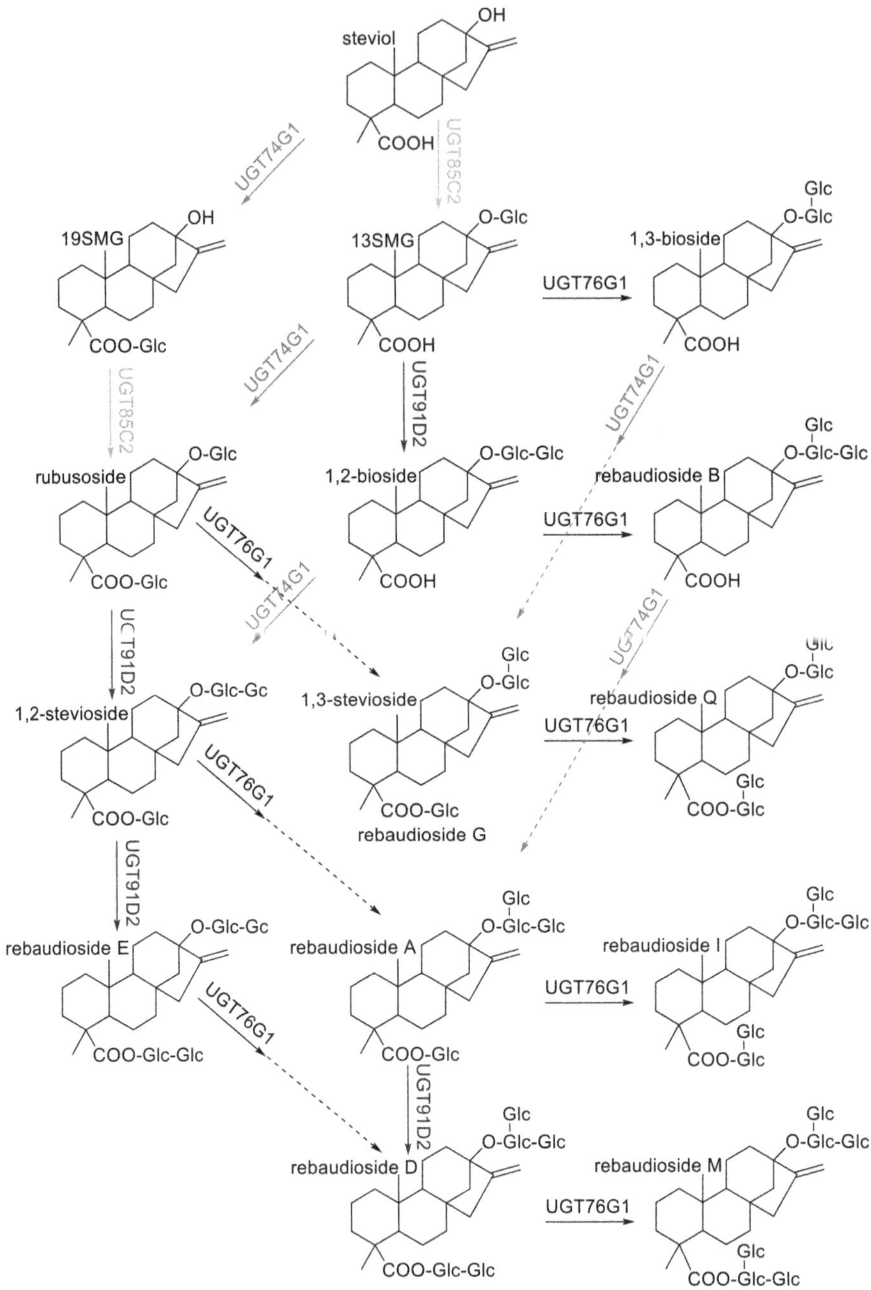

Figure 2.10 Metabolic grid of glucosylation reactions leading to numerous steviol glycosides. Glc, glucose; 13SMG, steviol-13-*O*-monoglucoside; 19SMG, steviol-19-*O*-monoglucose ester. Adapted with permission from ref. 134, https://doi.org/10.1186/s12934-016-0609-1, © The Authors 2016 under a CC BY 4.0 license, http://creativecommons.org/licenses/by/4.0/.

rise to a β-D-glucoside and a glucose ester. UGT91D2 and UGT76G1 elongate the sugars directly bound to the C_{13}- and C_{19}-position with additional D-glucose units *via* formation of 1,2-, and 1,3-β-D-glucosidic bonds, respectively. UGT76G1 catalyses the transfer of D-glucose to the 3-hydroxyl group on both mono- and 1,2-disaccharides attached to the C_{13} and C_{19} positions of steviol. In contrast, UGT91D2 catalyses the 1,2-β-D-glucosylation of monosaccharides present at the C_{13} and C_{19} position but not in 1,3-disaccharides. Few steviol glycosides are also decorated with L-rhamnose units but a rhamnosyltransferase has not been functionally characterised in *Stevia rebaudiana* until now.

2.3.2 MVA or DXP Pathway?

There are two different biosynthetic pathways contributing to the early steps of the terpene biosynthesis in plants. For a long time, the mevalonic acid (MVA) pathway was believed to be the only pathway involved in generating C_5 isoprene units, IPP and its isomer dimethylallyl diphosphate (DMAPP). However, another mevalonic acid independent pathway, the deoxyxylulose phosphate (DXP) or also called methylerythritol phosphate (MEP) pathway, was discovered in the early 1990s: firstly in bacteria (*Zymomonas mobilis*), later on in algae and higher plant chloroplasts as well.[5,87–91] Discovery of the DXP/MEP-pathway was made by inhibition experiments with mevinolin, which impedes formation of mevalonate but did not affect the formation of plastid isoprenoids, such as carotenoids and phytol (chlorophyll sidechain). This indicated that there must be a mevalonate independent pathway located in the plant's plastids.[5,92]

Furthermore, labelling experiments showed unexpected results. For example, the incorporation ratio of [2-^{14}C]-acetate and [1-^{14}C]-acetate, precursors of the MVA-pathway, into the tetraterpene lycopene, the red colour of tomatoes, or into the isoprenoid sidechain of ubiquinones produced by bacteria did not comply with incorporation found in related terpenoid structures in animal tissue like squalene and cholesterol, which are produced *via* the MVA-pathway.[93,94] Very illuminating were feeding experiments with ^{13}C-glucose. Feeding to *Zymomonas mobilis* bacteria proved that hopanoids, which are triterpenoid elements of the cell membrane, must derive from C_2-units synthesised *via* pyruvate decarboxylation, as revealed incorporation patterns of monoterpenes in young peppermint shoots (*Mentha x piperita*).[87,91,95] Thus, the glycolytic source of the DXP/MEP pathway was established.

In contrast, MVA is a product of the acetate metabolism and IPP and DMAPP are built *via* six steps (see Figure 2.11). A precursor of MVA is β-hydroxy-β-methylglutaryl-CoA (HMG-CoA), which is formed by the condensation of two acetyl-coenzyme A (AC-CoA) molecules catalysed by acetoacetyl (AcAc)-CoA thiolase (AACT). The addition of another enzyme bound acetyl-CoA, by the aid of HMG-CoA synthase (HMGS), gives the branched chain of HMG-CoA.[96] A two-step NADPH dependent reduction, catalysed by HMG-CoA reductase (HMGR) finally forms MVA from HMG-CoA.[97] ATP-dependent

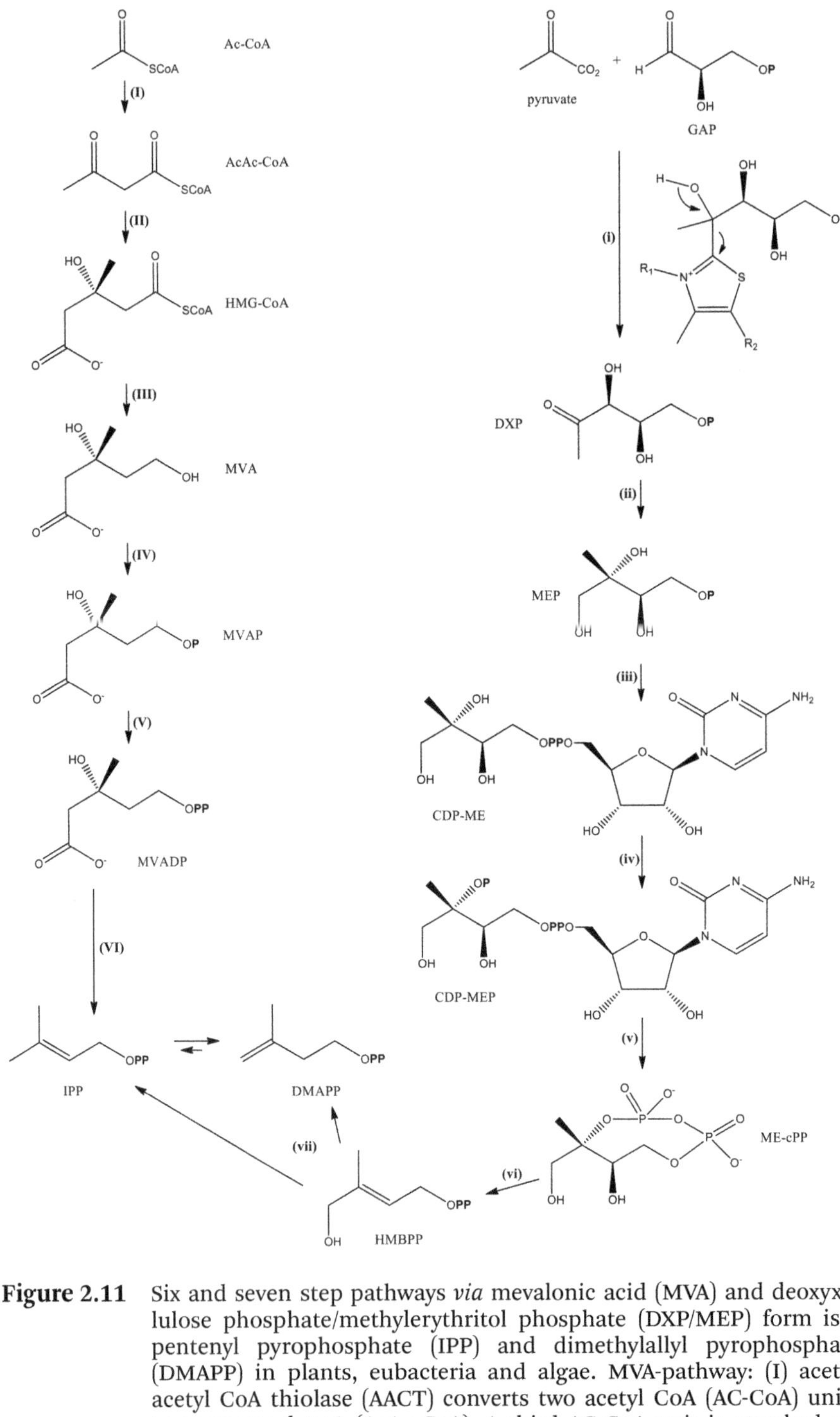

Figure 2.11 Six and seven step pathways *via* mevalonic acid (MVA) and deoxyxylulose phosphate/methylerythritol phosphate (DXP/MEP) form isopentenyl pyrophosphate (IPP) and dimethylallyl pyrophosphate (DMAPP) in plants, eubacteria and algae. MVA-pathway: (I) acetoacetyl CoA thiolase (AACT) converts two acetyl CoA (AC-CoA) units to acetoacetyl CoA (AcAc-CoA). A third AC-CoA unit is attached *via*

phosphorylation catalysed by MVA kinase (MK) and phospho-MVA kinase (PMK) results in mevalonic acid diphosphate (MVADP), which has a six carbon skeleton.[89,98,99] Diphospho-MVA decarboxylase (PPMD) catalyses the loss of CO_2 and IPP is formed. Finally, isomerase activity ultimately favours the production of DMAPP.[99,100]

1-Deoxy-D-5-xylulose 5-phosphate (DXP) is formed by the addition of pyruvic acid and glyceraldehyde 3-phosphate (GAP) (see Figure 2.11), which are, as mentioned above, intermediates of the glycolytic pathway. The addition is performed *via* a thiamine–diphosphate mediated enamine-complex and the loss of CO_2, catalysed by the enzyme DXP synthase (DXS).[101,102] A 1-deoxy-D-5-xylulose 5-phosphate reducto-isomerase (DXR) catalyses intramolecular rearrangement and NADPH mediated reduction, which eventually leads to 2-*C*-methyl-D-erythritol 4-phosphate (MEP).

To form IPP *via* the DXP/MEP-pathway a cascade of five further enzymatic reactions is required and was elucidated in the early years of the new millennium: a cytidyl derivate of MEP is built *via* a diphospho bridge by 4-(cytidine 5′-diphospho)-2-*C*-methyl-D-erythritol synthase (MCT), forming 4-(cytidine 5′-diphospho)-2-*C*-methyl-D-erythritol (CDP-ME).[89,103–106] This mechanism seems quite analogue to the activation of glucose by addition of uridine triphosphate (UTP), forming UDP-glucose. CDP-ME is further phosphorylated to 2-phospho-4-(cytidine 5′-diphospho)-2-*C*-methyl-D-erythritol (CDP-MEP) by 4-(cytidine 5′-diphospho)-2-*C*-methyl-D-erythritol kinase (CMK) and adenosine triphosphate (ATP) as substrate.[107] A cyclisation reaction follows and leads to 2-*C*-methyl-D-erythritol 2,4-cyclodiphosphate (ME-cPP), catalysed by the ME-cPP-synthase (MDS).[108] Reduction to 1-hydroxy-2-methyl-2-(*E*)-butenyl 4-diphosphate (HMBPP) is the penultimate step *via* catalysis by HMBPP-synthase (HDS).[109,110] Finally, HMBPP is converted to IPP and in smaller amounts to DMAPP, catalysed by HMBPP-reductase (HDR).[110–112]

(II) β-hydroxy-β-methylglutaryl-CoA synthase (HMGS). MVA is built by (III) β-hydroxy-β-methylglutaryl-CoA reductase (HMGR) catalysis and phosphorylated twice by (IV) MVA kinase (MK) and (V) phospho-MVA kinase (PMK). A carbon loss follows by (VI) diphospho-MVA decarboxylase (PPMD) catalysis and IPP is built. Isomerase finally favours the building of DMAPP. DXP/MEP-pathway: DXP is formed from pyruvate and glyceraldehyde 3-phosphate (GAP) by (i) DXP synthase (DXS). (ii) DXP reducto-isomerase (DXR) catalyses the two-step rearrangement and reduction to MEP, which is formed to 4-(cytidine 5′-diphospho)-2-*C*-methyl-D-erythritol (CDP-ME) by (iii) 4-(cytidine 5′-diphospho)-2-*C*-methyl-D-erythritol synthase (MCT) and further on to 2-phospho-4-(cytidine 5′-diphospho)-2-*C*-methyl-D-erythritol (CDP-MEP) by (iv) 4-(cytidine 5′-diphospho)-2-*C*-methyl-D-erythritol kinase (CMK) catalysis. (v) 2-*C*-methyl-D-erythritol 2,4-cyclodiphosphate synthase (MDS) catalyses the building of 2-*C*-methyl-D-erythritol 2,4-cyclodiphosphate (ME-cPP), which is reduced to 1-hydroxy-2-methyl-2-(*E*)-butenyl 4-diphosphate (HMBPP) by (vi) HMBPP-synthase (HDS). Lastly, (vii) HMBPP-reductase (HDR) catalyses the formation of IPP and DMAPP.

While the MVA-pathway is located in the cytosol of plant, bacteria and animal cells, DXP mediated isoprene synthesis takes place in the plastids of plants, algae and bacteria tissue but not in animal cells.[89]

Furthermore, several studies on the formation of terpenoids indicate, that the MVA-pathway mainly contributes to the formation of monoterpenes (C_5) and diterpenes (C_{20}), while the DXP-pathway more likely leads to sesquiterpenes (C_{15}) and triterpenes (C_{30}), yet very dependent on the plant family.[89]

Probably one of the reasons for the late discovery of the DXP/MEP pathway is the metabolic cross talk of both pathways, as low incorporation rates in feeding experiments into plant terpenoids at first were believed to be due to restricted uptake of the labelled MVA precursors.[113] Many studies gave evidence for the cross talk between the MVA and DXP/MEP-pathway, *e.g.* incorporation of [1,1,1,4-$^2H_4$]-deoxy-D-xylulose (DOX) into sterols of tobacco plant cells[114] or the even distribution of incorporated [5,5-$^2H^2$]-mevalonic acid lactone (d_2-MVL) and [5,5-$^2H_2$]-DOX into β-caryophyllene of carrot roots (*Daucus carota* L.).[115]

Hydroponic feeding experiments of stevia shoot tips with [1-^{13}C]-glucose or deuterium labelled [5,5-$^2H_2$]-DOX (d_2-DOX) revealed that isoprenes originating from the DXP-pathway are incorporated into steviol glycosides.[3,4] Although, same experiments with old leaves prove that d_2-MVL is incorporated into the precursor *ent* kaurene with up to 15%, incorporation in steviol glycosides could not be determined. It was further shown that the DXP/MEP-pathway most likely contributes to the formation of monoterpenes, whereas cross talk of the MVA- and DXP/MEP-pathway was documented for some sesquiterpenes in stevia.[4]

The stevia DXS and DXR genes were characterised just shortly after the discovery of the DXP/MEP-pathway,[116] whereas the genes encoding the following enzymes of the DXP/MEP-pathway, namely MCT, CMK, MDS, HDS and HDR were just recently cloned and their expression pattern analysed.[7,76]

2.3.3 Site of Biosynthesis

There are several analytical methodologies to predict the cellular localisation of enzymes and thus to find out where intermediates of biosynthetic pathways are formed. Modern techniques of localizing enzymes focus upon investigation of the amino acid sequences of enzymes in combination with algorithmic programs like PSORT, TargetP, ChloroP and others, which allow predicting the location of enzymes.[117–120] Another quite modern technique involves the fusion of the desired enzymes with fluorescent substrates, for example the green fluorescent protein (GFP), which is naturally occurring *e.g.* in the jellyfish *Aequorea victoria*,[121,122] making enzymes microscopically visible by fluorescence.[123,124]

Figure 2.12 gives a schematic overview about the localisation of the biosynthetic pathway of steviol glycosides in the plant cell. Enzymes of the early steps of the steviol glycoside biosynthesis, *i.e.* the DXP/MEP-pathway, are believed to be located in the chloroplast and many studies, not only with *Stevia rebaudiana* plants, support this assumption.

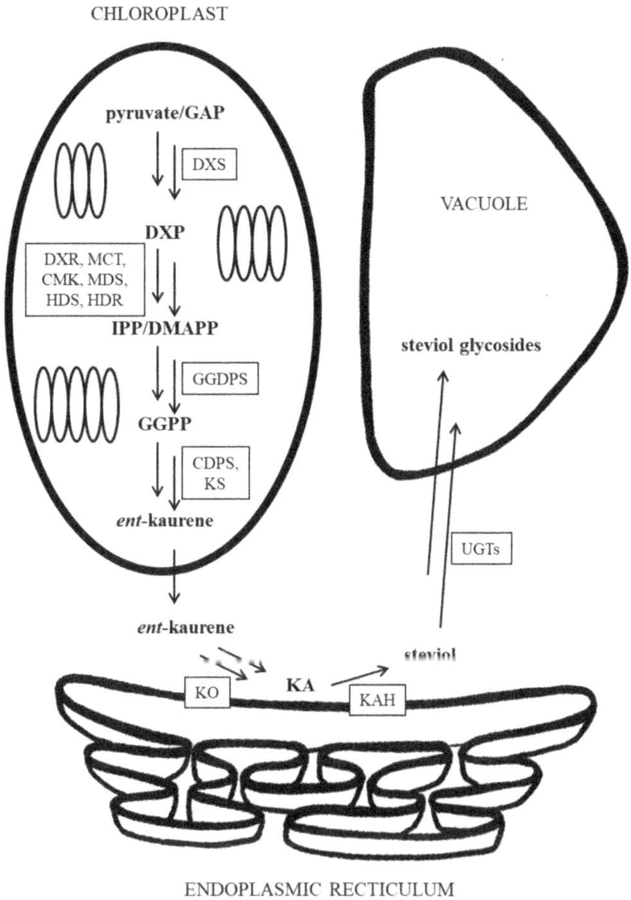

Figure 2.12 Cellular localisation of the biosynthetic steps which lead to the formation of steviol glycosides. Pyruvate and glyceraldehyde 3-phosphate (GAP) form isopentenyl pyrophosphate (IPP) and dimethylallyl pyrophosphate (DMAPP) *via* the deoxyxylulose phosphate/methylerythritol phosphate (DXP/MEP)-pathway (DXS; DXP synthase, DXR; DXP reducto-isomerase, MCT; 4-(cytidine 5′-diphospho)-2-*C*-methyl-D-erythritol synthase, CMK; 4-(cytidine 5′-diphospho)-2-*C*-methyl-D-erythritol kinase, MDS; 2-*C*-methyl-D-erythritol 2,4-cyclodiphosphate synthase, HDS; 1-hydroxy-2-methyl-2-(*E*)-butenyl 4-diphosphate-synthase and HDR; 1-hydroxy-2-methyl-2-(*E*)-butenyl 4-diphosphate-reductase) in the chloroplasts. IPP/DMAPP further form geranylgeranyl pyrophosphate (GGPP) (GGDPS; geranylgeranyl pyrophosphate-synthase) and eventually *ent*-kaurene (CDPS; copalyl diphosphate synthase and KS; kaurene synthase). Enzymes responsible are located in the plastids as well. *Ent*-kaurene is then transported to the cytosol, where it is oxidised to steviol by cytochrome P450 mono-oxygenases (KO; kaurene oxidase and KAH; kaurenoic acid 13-hydroxylase), located at the endoplasmic reticulum (ER). Glycosylation of steviol is performed by uridine diphosphate-glycosyltransferases (UGTs) in the cytosol and steviol glycosides are finally stored in the vacuoles.

Plastidial targeting sequences of DXS and DXR enzymes in stevia affirm the synthesis of DXP and MEP in the chloroplasts of stevia plant cells.[116] These findings correlate with former investigations of DXS and DXR in other plant families, *e.g.* Lamiaceae.[6,101]

MCT was found to have a chloroplast transit peptide sequence as well, evaluated by ChloroP, *i.e.* the protein is transported from the cytosol into the chloroplast. However, these results were obtained from studies with *Ginkgo biloba*.[125] GFP-fusion experiments of MCT, CMK, MDS, HDS and HDR in *Arabidopsis thaliana* protoplasts showed fluorescence in the chloroplasts, intensively substantiating that the DXP/MEP-pathway is a plastidial pathway.[126] The following condensation and cyclisation steps seem to take place in the plastid as well, as GGDPS, CDPS and KS contain chloroplast localisation signals in *Stevia rebaudiana*.[76]

Ent-kaurene, the final product of KS, is then transported into the cytosol, where two cytochrome P450 mono-oxygenases finally oxygenate *ent*-kaurene to steviol. A chloroplastic KO, attached to the outer membrane of the chloroplast, as it was reported previously for the related gibberellin pathway in *Arabidopsis*, might mediate the transport of the lipophilic *ent*-kaurene into the cytosol.[127]

The subcellular location prediction program PSORT and expression analyses of green fluorescent protein (GFP) fusions with KO, however, revealed that the fused protein is attached to the endoplasmic reticulum only in *Stevia rebaudiana*. The same location is proposed for KAH.[80]

Different UGT-GFPs in contrast, showed a fluorescence pattern characteristic for cytoplasmic localisation.[80] Using PSORT to predict the subcellular localisation showed unclear results. It is therefore most probable that the enzymes are cytoplasmic.[84]

Storage of the steviol glycosides eventually is assumed to be linked to the vacuoles due to their water solubility.[13]

2.4 HS-SPME-GC-MS and LC-MS/MS-analysis of Foliar Trichomes and Exudates

Analytical investigations of morphological plant structures enabled first insights into the sites where steviol glycosides and precursors are produced and stored. Many chemical analyses *via* liquid and gas chromatography, of whole stevia leaves and extracts, have been performed so far,[128–131] but only few investigations have focused upon glandular trichomes of stevia, although there is some evidence that trichome density might be linked to a higher steviol glycoside content.[28]

2.4.1 HS-SPME-GC-MS

Lately, head space-solid phase micro extraction (HS-SPME)-GC-MS measurements of whole stevia leaves revealed the presence of *ent*-kaurene, a diterpenoid intermediate of the steviol glycoside biosynthesis. Further on, the ratio of labelled *ent*-kaurene to genuine *ent*-kaurene in foliar trichomes was more than 2.5 in hydroponical feeding experiments with d_2-DOX, indicating that *ent*-kaurene might be synthesised in trichome cells.[4]

However, recent comparative HS-SPME-GC-MS measurements, inspired by the described experiments,[4] of five different *Stevia rebaudiana* genotypes showed that *ent*-kaurene is detectable in the headspace of only one of the analysed genotypes, namely Gawi (see Figures 2.13 and 2.14). In addition, trichomes of ecotype Gawi were isolated by brushing them off adaxial and abaxial leaf surfaces as described in an earlier publication[56] and HS-SPME-GC-MS analysis of them showed a comparable terpene pattern to those of leaf HS-SPME-GC-MS analyses, indicating that leaf volatiles directly originate from the trichomes on the leaf surface. Interestingly, detection of *ent*-kaurene came along with an intensive peak for *epi*-manoyl oxide, again only detectable in genotype Gawi.[132]

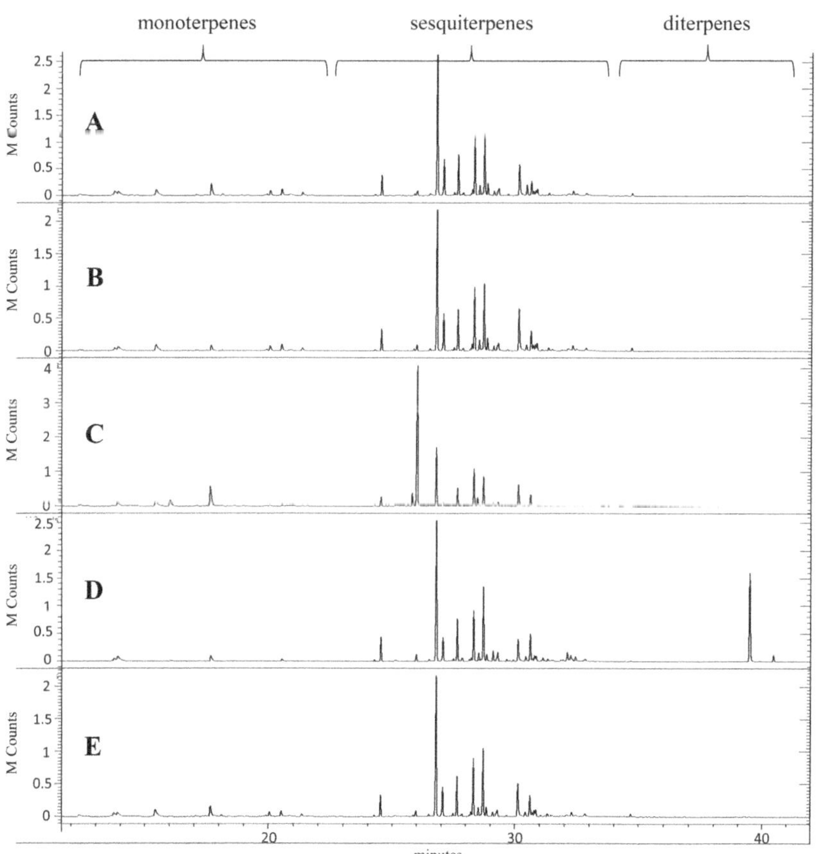

Figure 2.13 Comparative HS-SPME-GC-MS measurements of several stevia genotypes (A), (B), (C), (D) and (E) show differences in the composition of volatiles.

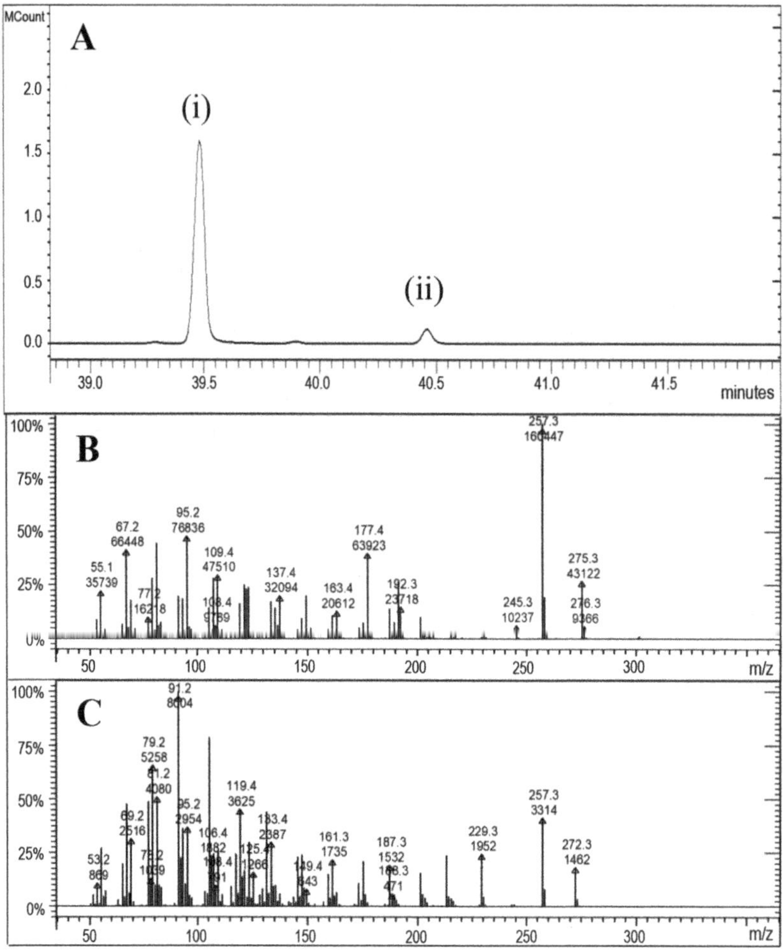

Figure 2.14 Enlarged GC-MS-chromatogram (A) with manoyl oxide (i) and *ent*-kaurene (ii) peaks and associated MS-spectra of manoyl oxide (B) and *ent*-kaurene (C), measuring a stevia leaf of genotype Gawi *via* HS-SPME.

2.4.2 LC-MS/MS

To substantiate GC-MS-measurements, which gave evidence that volatile precursors of the steviol glycoside synthesis are present in HS-SPME analysis, analyses were performed to check whether the non-volatile steviol glycosides are also detectable in trichome cells. These investigations were performed using a recently developed LC-MS/MS-method.[133] Again, trichomes were isolated by brushing them off adaxial and abaxial leaf surfaces as described in an earlier publication,[56] diluted in acetonitrile and water and filtered through a cellulose membrane. Further on, dip extractions with ethyl acetate and acetonitrile/water were performed to compare the chemical composition with regard to steviol glycosides of trichome exudates with those of isolated trichomes. Both extraction techniques, brushing trichomes off the leaf surface and dipping whole leaves for 10 s in an adequate solvent, revealed that steviol glycosides are present in trichome cells and exudates of the investigated genotype. Furthermore, the pattern of steviol glycosides was comparable to that found in extracts of whole leaves (data not shown).

2.5 Outlook

Recent innovations in bioengineering are focused on targeted production of steviol glycosides (in particular Reb D and Reb M) *via* genetically modified yeast, using *e.g.* corn dextrose as substrate.[2] However, it is questionable whether those production methods meet customer demand for a natural alternative to the other non-caloric but artificial sweeteners.

In this context, the detection of steviol glycosides in stevia trichomes, as shown above, may pave the way for future targeted natural breeding programs to alter the density and chemical composition of glandular trichomes in stevia plants of *e.g.* the Gawi ecotype.

References

1. M. I. Espinoza, J. P. Vincken, M. Sanders, C. Castro, M. Stieger and E. Agosin, *J. Agric. Food Chem.*, 2014, **62**(49), 11797.
2. U.S. Food and Drug Administration, 2016 (GRN No. 626).
3. N. Totté, L. Charon, M. Rohmer, F. Compernolle, I. Baboeuf and J. M. C. Geuns, *Tetrahedron Lett.*, 2000, **41**(33), 6407.
4. U. Wölwer-Rieck, B. May, C. Lankes and M. Wüst, *J. Agric. Food Chem.*, 2014, **62**(11), 2428.
5. H. K. Lichtenthaler, J. Schwender, A. Disch and M. Rohmer, *FEBS Lett.*, 1997, **400**(3), 271.
6. B. M. Lange and R. Croteau, *Arch. Biochem. Biophys.*, 1999, **365**(1), 170.
7. J. E. Brandle and P. G. Telmer, *Phytochemistry*, 2007, **68**(14), 1855.
8. A. S. Richman, M. Gijzen, A. N. Starratt, Z. Yang and J. E. Brandle, *Plant J.*, 1999, **19**(4), 411.
9. P. Hedden and A. L. Phillips, *Trends Plant Sci.*, 2000, **5**(12), 523.
10. W. Rademacher, *Annu. Rev. Plant Physiol. Plant Mol. Biol.*, 2000, **51**, 501.
11. M. Ruddat, A. Lang and E. Mosettig, *Naturwissenschaften*, 1963, **50**(1), 23.
12. S. Ceunen and J. M. C. Geuns, *J. Nat. Prod.*, 2013, **76**(6), 1201.
13. H. Shibata, Y. Sawa, T. Oka, S. Sonoke, K. K. Kim and M. Yoshioka, *Arch. Biochem. Biophys.*, 1995, **321**(2), 390.
14. N. P. D. Nanayakkara, J. A. Klocke, C. M. Compadre, R. A. Hussain, J. M. Pezzuto and A. D. Kinghorn, *J. Nat. Prod.*, 1987, **50**(3), 434.
15. R. de Guzman, PhD thesis, Central Queensland University, 2010.
16. S. Stoyanova, J. Geuns, E. Hideg and W. van den Ende, *Int. J. Food Sci. Nutr.*, 2011, **62**(3), 207.

17. S. Hajihashemi and J. M. C. Geuns, *Free Radicals Antioxid.*, 2013, **3**, 34.
18. F. Tateo, M. Bononi, M. G. Mariotti, L. Cornara and G. Serrato-Valenti, *Plant Biosyst.*, 2001, **135**(1), 25.
19. J. Gershenzon, M. Maffei and R. Croteau, *Plant Physiol.*, 1989, **89**(4), 1351.
20. D. McCaskill and R. Croteau, *Planta*, 1995, **197**(1), 49–56.
21. A. L. Schilmiller, I. Schauvinhold, M. Larson, R. Xu, A. L. Charbonneau, A. Schmidt, C. Wilkerson, R. L. Last and E. Pichersky, *Proc. Natl. Acad. Sci. U. S. A.*, 2009, **106**(26), 10865.
22. K. Brückner, D. Božić, D. Manzano, D. Papaefthimiou, I. Pateraki, U. Scheler, A. Ferrer, R. C. H. de Vos, A. K. Kanellis and A. Tissier, *Phytochemistry*, 2014, **101**, 52.
23. A. Huchelmann, M. Boutry and C. Hachez, *Plant Physiol.*, 2017, **175**(1), 6.
24. U.S. Food and Drug Administration, 2016 (GRN No. 619).
25. EU, Off. J. Europ. Union, 2012, L83, 1.
26. EU, Off. J. Europ. Union, 2011, L295, 1.
27. S. Ceunen and J. M. C. Geuns, *Plant Sci.*, 2013, **198**, 72.
28. N. I. Bondarev, M. A. Sukhanova, G. A. Semenova, O. V. Goryaeva, S. E. Andreeva and A. M. Nosov, *Moscow Univ. Biol. Sci. Bull.*, 2010, **65**(1), 12.
29. M. J. Kim, J. Jin, J. Zheng, L. Wong, N.-H. Chua and I.-C. Jang, *Plant Physiol.*, 2015, **169**(4), 2462.
30. D. D. Soejarto, in *Stevia*, ed. A. D. Kinghorn, Taylor & Francis, London, 2002, p. 18.
31. A. Singh, K. Singh, P. Singh and M. P. Singh, *Int. J. Adv. Res.*, 2015, 628.
32. G. Turner, J. Gershenzon, E. E. Nielson, J. E. Froehlich and R. Croteau, *Plant Physiol.*, 1999, **120**(3), 879.
33. E. Werker, in *Advances in Botanical Research: Incorporation Advances in Plant Pathology Plant Trichomes*, ed. D. L. Hallahan, J. C. Gray and J. A. Callow, Academic Press, San Diego, CA, 2000, p. 1.
34. B. M. Lange and G. W. Turner, *Plant Biotechnol. J.*, 2013, **11**(1), 2.
35. R. Baur, S. Binder and G. Benz, *Oecologia*, 1991, **87**(2), 219.
36. D. Tian, J. Tooker, M. Peiffer, S. H. Chung and G. W. Felton, *Planta*, 2012, **236**(4), 1053.
37. A. Fahn, in *Advances in Botanical Research: Incorporation Advances in Plant Pathology Plant Trichomes*, ed. D. L. Hallahan, J. C. Gray and J. A. Callow, Academic Press, San Diego, CA, 2000, p. 37.
38. G. J. Wagner, E. Wang and R. W. Shepherd, *Ann. Bot.*, 2004, **93**(1), 3.
39. J. J. Glas, B. C. J. Schimmel, J. M. Alba, R. Escobar-Bravo, R. C. Schuurink and M. R. Kant, *Int. J. Mol. Sci.*, 2012, **13**(12), 17077.
40. W. R. Monteiro, M. de Moraescastro, S. C. Mazzoni-Viveiros and P. G. Mahlberg, *Braz. J. Bot.*, 2001, **24**(3), 349.
41. W. R. Monteiro, W. Caldeira, M. d. M. Castro, P. G. Mahlberg, M. V. Cruz and C. H. Bornman, *S. Afr. J. Bot.*, 2003, **69**(4), 489.
42. N. I. Bondarev, M. A. Sukhanova, O. V. Reshetnyak and A. M. Nosov, *Biol. Plant.*, 2004, **47**(2), 261.

43. P. Guleria, V. Kumar and S. K. Yadav, *Asian J. Plant Sci.*, 2011, **10**(8), 401.
44. D. R. Gang, *Plant Physiol.*, 2001, **125**(2), 539.
45. A. Schilmiller, F. Shi, J. Kim, A. L. Charbonneau, D. Holmes, A. Daniel Jones and R. L. Last, *Plant J.*, 2010, **62**(3), 391.
46. C. Sallaud, C. Giacalone, R. Töpfer, S. Goepfert, N. Bakaher, S. Rösti and A. Tissier, *Plant J.*, 2012, **72**(1), 1.
47. R. Croteau, *Plant Physiol.*, 1977, **59**(3), 519.
48. M. Maffei, F. Chialva and T. Sacco, *New Phytol.*, 1989, **111**(4), 707.
49. N. Aziz, N. L. Paiva, G. D. May and R. A. Dixon, *Planta*, 2005, **221**(1), 28.
50. H. Cui, S.-T. Zhang, H.-J. Yang, H. Ji and X.-J. Wang, *BMC Plant Biol.*, 2011, **11**, 76.
51. W. Wang, Y. Wang, Q. Zhang, Y. Qi and D. Guo, *BMC Genomics*, 2009, **10**, 465.
52. S. S. A. Soetaert, C. M. F. van Neste, M. L. Vandewoestyne, S. R. Head, A. Goossens, F. C. W. van Nieuwerburgh and D. L. D. Deforce, *BMC Plant Biol.*, 2013, **13**, 220.
53. B. Ghosh, T. C. Westbrook and A. D. Jones, *Metabolomics*, 2014, **10**(3), 496.
54. J. C. Snyder and J. P. Hyatt, *Plant Sci. Lett.*, 1984, **37**(1–2), 177.
55. J. H. Kang, G. Liu, F. Shi, A. D. Jones, R. M. Beaudry and G. A. Howe, *Plant Physiol.*, 2010, **154**(1), 262.
56. E. A. P. Ekanayaka, C. Li and A. D. Jones, *Phytochemistry*, 2014, **98**, 223.
57. R. Croteau and J. N. Winters, *Plant Physiol.*, 1982, **69**(4), 975.
58. C. K. Keene and G. J. Wagner, *Plant Physiol.*, 1985, **79**(4), 1026.
59. E. Wang, R. Wang, J. DeParasis, J. H. Loughrin, S. Gan and G. J. Wagner, *Nat. Biotechnol.*, 2001, **19**(4), 371.
60. X. Zhang and D. G. Oppenheimer, *Plant Cell Physiol.*, 2004, **45**(2), 221.
61. E. Harada, J.-A. Kim, A. J. Meyer, R. Hell, S. Clemens and Y.-E. Choi, *Plant Cell Physiol.*, 2010, **51**(10), 1627.
62. G. U. Balcke, S. Bennewitz, S. Zabel and A. Tissier, in *Plant Isoprenoids*, ed. M. Rodríguez-Concepción, Humana Press, New York, 2014.
63. T. Yamaura, S. Tanaka and M. Tabata, *Planta Med.*, 1992, **58**(2), 153.
64. J. Gershenzon, M. A. Duffy, F. Karp and R. Croteau, *Anal. Biochem.*, 1987, **163**(1), 159.
65. J. Gershenzon, D. McCaskill, J. I. M. Rajaonarivony, C. Mihaliak, F. Karp and R. Croteau, *Anal. Biochem.*, 1992, **200**(1), 130.
66. B. M. Lange, M. R. Wildung, E. J. Stauber, C. Sanchez, D. Pouchnik and R. Croteau, *Proc. Natl. Acad. Sci. U. S. A.*, 2000, **97**(6), 2934.
67. M. E. Olsson, L. M. Olofsson, A.-L. Lindahl, A. Lundgren, M. Brodelius and P. E. Brodelius, *Phytochemistry*, 2009, **70**(9), 1123.
68. L. Olofsson, A. Lundgren and P. E. Brodelius, *Plant Sci.*, 2012, **183**, 9.
69. M. Majdi, Q. Liu, G. Karimzadeh, M. A. Malboobi, J. Beekwilder, K. Cankar, R. d. Vos, S. Todorović, A. Simonović and H. Bouwmeester, *Phytochemistry*, 2011, **72**(14–15), 1739.
70. A. Sallets, M. Beyaert, M. Boutry and A. Champagne, *J. Proteome Res.*, 2014, **13**(7), 3386.

71. P. Grassi, J. Novak, H. Steinlesberger and C. Franz, *Phytochem. Anal.*, 2004, **15**(3), 198.
72. J.-H. Kang, G. Liu, F. Shi, A. D. Jones, R. M. Beaudry and G. A. Howe, *Plant Physiol.*, 2010, **154**(1), 262.
73. A. Oehme and M. Wüst, *Lebensmittelchemie*, 2016, **70**, 107.
74. G. Wang, L. Tian, N. Aziz, P. Broun, X. Dai, J. He, A. King, P. X. Zhao and R. A. Dixon, *Plant Physiol.*, 2008, **148**(3), 1254.
75. S. Purkayastha, A. Markosyan, I. Prakash, S. Bhusari, G. Pugh, B. Lynch and A. Roberts, *Regul. Toxicol. Pharmacol.*, 2016, **77**, 125.
76. H. Kumar, K. Kaul, S. Bajpai-Gupta, V. K. Kaul and S. Kumar, *Gene*, 2012, **492**(1), 276.
77. P. M. Dewick, in *Medicinal Natural Products: A Biosynthetic Approach*, ed. P. M. Dewick, John Wiley & Sons, Chichester, 2008, p. 167.
78. S. Ceunen, PhD thesis, Katholieke Universiteit Leuven, 2012.
79. J. MacMillan, *Nat. Prod. Rep.*, 1997, **14**(3), 221.
80. T. V. Humphrey, A. S. Richman, R. Menassa and J. E. Brandle, *Plant Mol. Biol.*, 2006, **61**(1–2), 47.
81. C. A. Helliwell, A. Poole, W. James Peacock and E. S. Dennis, *Plant Physiol.*, 1999, **119**(2), 507.
82. K. K. Kim, Y. Sawa and H. Shibata, *Arch. Biochem. Biophys.*, 1996, **332**(2), 223.
83. J. Brandle and A. Richman, *US Pat.* 7927,851 B2, 2011.
84. A. Richman, A. Swanson, T. Humphrey, R. Chapman, B. McGarvey, R. Pocs and J. Brandle, *Plant J.*, 2005, **41**(1), 56.
85. H. Madhav, S. Bhasker and M. Chinnamma, *Plant Physiol. Biochem.*, 2013, **63**, 245.
86. A. A. A. Mohamed, S. Ceunen, J. M. C. Geuns, W. van den Ende and M. de Ley, *J. Plant Physiol.*, 2011, **168**(10), 1136.
87. M. Rohmer, M. Knani, P. Simonin, B. Sutter and H. Sahm, *Biochem. J.*, 1993, **295**(2), 517.
88. M. Rohmer, *Nat. Prod. Rep.*, 1999, **16**(5), 565.
89. A. Hemmerlin, J. L. Harwood and T. J. Bach, *Prog. Lipid Res.*, 2012, **51**(2), 95.
90. S. Horbach, H. Sahm and R. Welle, *FEMS Microbiol. Lett.*, 1993, **111**(2–3), 135.
91. M. Rohmer, *Pure Appl. Chem.*, 2003, **75**(2–3), 37.
92. T. J. Bach and H. K. Lichtenthaler, *Physiol. Plant.*, 1983, **59**(1), 50.
93. I. Zabin, *J. Biol. Chem.*, 1957, **226**, 851.
94. S. Pandian, S. Saengchjan and T. S. Raman, *Biochem. J.*, 1981, **196**(3), 675.
95. W. Eisenreich, S. Sagner, M. H. Zenk and A. Bacher, *Tetrahedron Lett.*, 1997, **38**(22), 3889.
96. H. Rudney and J. J. Ferguson, *J. Am. Chem. Soc.*, 1957, **79**(20), 5580.
97. J. J. Ferguson, I. F. Durr and H. Rudney, *Proc. Natl. Acad. Sci. U. S. A.*, 1959, **45**(4), 499.
98. T. T. Tchen, *J. Biol. Chem.*, 1958, **233**(5), 1100.
99. K. Bloch, S. Chaykin, A. H. Phillips and A. De Waard, *J. Biol. Chem.*, 1959, **234**(10), 2595.

100. B. W. Agranoff, H. Eggerer, U. Henning and F. Lynen, *J. Biol. Chem.*, 1960, **235**, 326.
101. B. M. Lange, M. R. Wildung, D. McCaskill and R. Croteau, *Proc. Natl. Acad. Sci. U. S. A.*, 1998, **95**(5), 2100.
102. G. A. Sprenger, U. Schörken, T. Wiegert, S. Grolle, A. A. de Graaf, S. V. Taylor, T. P. Begley, S. Bringer-Meyer and H. Sahm, *Proc. Natl. Acad. Sci. U. S. A.*, 1997, **94**(24), 12857.
103. E. Cordoba, M. Salmi and P. León, *J. Exp. Bot.*, 2009, **60**(10), 2933.
104. M. A. Phillips, P. León, A. Boronat and M. Rodríguez-Concepción, *Trends Plant Sci.*, 2008, **13**(12), 619.
105. T. Kuzuyama, M. Takagi, K. Kaneda, T. Dairi and H. Seto, *Tetrahedron Lett.*, 2000, **41**(5), 703.
106. F. Rohdich, J. Wungsintaweekul, M. Fellermeier, S. Sagner, S. Herz, K. Kis, W. Eisenreich, A. Bacher and M. H. Zenk, *Proc. Natl. Acad. Sci. U. S. A.*, 1999, **96**(21), 11758.
107. T. Kuzuyama, M. Takagi, K. Kaneda, H. Watanabe, T. Dairi and H. Seto, *Tetrahedron Lett.*, 2000, **41**(16), 2925.
108. M. Takagi, T. Kuzuyama, K. Kaneda, H. Watanabe, T. Dairi and H. Seto, *Tetrahedron Lett.*, 2000, **41**(18), 3395.
109. S.-M. Kim and S.-U. Kim, *Mol. Biol. Rep.*, 2010, **37**(2), 973.
110. W. N. Hunter, *J. Biol. Chem.*, 2007, **282**(30), 21573.
111. S.-M. Kim, T. Kuzuyama, A. Kobayashi, T. Sando, Y.-J. Chang and S.-U. Kim, *Planta*, 2008, **227**(2), 287.
112. M. Seemann, N. Campos, M. Rodríguez-Concepción, J.-F. Hoeffler, C. Grosdemange-Billiard, A. Boronat and M. Rohmer, *Tetrahedron Lett.*, 2002, **43**(5), 775.
113. W. Eisenreich, F. Rohdich and A. Bacher, *Trends Plant Sci.*, 2001, **6**(2), 78.
114. A. Hemmerlin, J.-F. Hoeffler, O. Meyer, D. Tritsch, I. A. Kagan, C. Grosdemange-Billiard, M. Rohmer and T. J. Bach, *J. Biol. Chem.*, 2003, **278**(29), 26666.
115. D. Hampel, A. Mosandl and M. Wüst, *Phytochemistry*, 2005, **66**(3), 305.
116. N. Totté, W. van den Ende, E. J. M. van Damme, F. Compernolle, I. Baboeuf and J. M. C. Geuns, *Can. J. Bot.*, 2003, **81**(5), 517.
117. P. Horton, K.-J. Park, T. Obayashi, N. Fujita, H. Harada, C. J. Adams-Collier and K. Nakai, *Nucleic Acids Res.*, 2007, **35**(Web Server issue), W585–W587.
118. K. Nakai and P. Horton, *Trends Biochem. Sci.*, 1999, **24**(1), 34.
119. O. Emanuelsson, H. Nielsen and G. von Heijne, *Protein Sci.*, 1999, **8**(5), 978.
120. O. Emanuelsson, S. Brunak, G. von Heijne and H. Nielsen, *Nat. Protoc.*, 2007, **2**(4), 953.
121. O. Shimomura, F. H. Johnson and Y. Saiga, *J. Cell. Comp. Physiol.*, 1962, **59**(3), 223.
122. D. C. Prasher, V. K. Eckenrode, W. W. Ward, F. G. Prendergast and M. J. Cormier, *Gene*, 1992, **111**(2), 229.
123. M. Chalfie, Y. Tu, G. Euskirchen, W. Ward and D. Prasher, *Science*, 1994, **263**(5148), 802.

124. R. Y. Tsien, *Annu. Rev. Biochem.*, 1998, **67**, 509.
125. S.-M. Kim, T. Kuzuyama, Y.-J. Chang, H.-J. Kwon and S.-U. Kim, *Phytochemistry*, 2006, **67**(14), 1435.
126. M.-H. Hsieh, C.-Y. Chang, S.-J. Hsu and J.-J. Chen, *Plant Mol. Biol.*, 2008, **66**(6), 663.
127. C. A. Helliwell, J. A. Sullivan, R. M. Mould, J. C. Gray, W. J. Peacock and E. S. Dennis, *Plant J.*, 2001, **28**(2), 201.
128. M. Tadhani and R. Subhash, *J. Med. Sci.*, 2006, **6**(3), 321.
129. U. Woelwer-Rieck, C. Lankes, A. Wawrzun and M. Wüst, *Eur. Food Res. Technol.*, 2010, **231**(4), 581.
130. U. Wölwer-Rieck, *J. Agric. Food Chem.*, 2012, **60**(4), 886.
131. K. M. Pierce, J. L. Hope, J. C. Hoggard and R. E. Synovec, *Talanta*, 2006, **70**(4), 797.
132. A. Oehme and M. Wüst, in *45. Deutscher Lebensmittelchemikertag*, Gesellschaft Deutscher Chemiker e.V., Frankfurt am Main, 2016, p. 72.
133. A. Oehme, M. Wüst and U. Wölwer-Rieck, *Int. J. Food Sci. Technol.*, 2017, **52**(10), 2156.
134. K. Olsson, S. Carlsen, A. Semmler, E. Simón, M. D. Mikkelsen and B. L. Møller, *Microb. Cell Fact.*, 2016, **15**(1), 207.

CHAPTER 3

Steviol Glycosides Production: Traditional Versus *New Technologies*

J. PERRET*

Stevia Natura SAS, 5 rue Georges Bizet, 63200 Riom, France
*E-mail: joelperret@stevia-natura.fr

3.1 Introduction

Stevia rebaudiana has been used for centuries by the Guarani's local tribes as a sweetener for their bitter *maté* tea or as a medicinal plant.[41] *Stevia rebaudiana* is one of the very few plants species to accumulate sweet molecules identified as diterpenoids glycosides: the steviol glycosides, earlier known as steviosides. Within this family more than 40 different steviol glycosides have been identified. Ten of these steviol glycosides are easily detectable in the native plant leaves with a major part represented by stevioside and rebaudioside A.[14,29,41,42,54,110]

Today, steviol glycoside sweeteners (extracted from *Stevia rebaudiana*) have an increasing consumer interest due to their natural origin, sustainability and environmental concerns.

Steviol glycosides are a promising alternative to synthetic sweeteners; since their authorisation in the United States and Switzerland in 2008, in France in

Food Chemistry, Function and Analysis No. 7
Steviol Glycosides: Cultivation, Processing, Analysis and Applications in Food
Edited by Ursula Wölwer-Rieck

Published by the Royal Society of Chemistry, www.rsc.org

2009, and then in the European Union in 2011. This market is growing and food companies introduce more and more steviol glycosides in their formulations for added sugar reduction.

Some weakness has been attributed to steviol glycosides mainly concerning taste issues, such as: a liquorice taste, off-flavours and lingering. Are these characteristics linked to steviol glycosides in general, or only to some specific steviol glycosides or to the so-called impurities left in the end product? Of course, the extraction and purification of these molecules must have an impact on the final taste quality of the product.[26]

The extraction processes are also of crucial importance in the definition of a natural product, so the technologies to be used for the extraction and purification of steviol glycosides have to take into account regulation constraints; most of the technologies developed by the industry have these constraints behind their objectives.

Many different technologies for the purification of *Stevia rebaudiana* extracts with different levels of purity have been developed and published in the literature. Industrial extraction started in Japan in the 1970s. The evolution of these technologies is governed by different aims: scientific or industrial/economic reasons, but also by regulation. For example, in the US the admission of steviol glycosides as a sweetener in food was first of all given only to purified rebaudioside A (rebaudioside A GRAS status).[94,95] This implied that this steviol glycoside had to be purified alone and that the technology to do it had to be developed. Or perhaps as the technology became available the GRAS assessment was written accordingly. Regulation, however, has a strong impact on technology.

In this chapter, different methods and processes used for steviol glycosides purification according to the final use of the product, history and major achievements will be discussed. It is important to distinguish between the work done with a scientific aim and the processes developed for an industrial production as it is clear that what could be achieved in a laboratory is not necessarily up scalable to an industrial process line in terms of efficiency, yield, quality or cost.

New technologies, like biotechnology, as well as their influence on new products and on regulation will also be discussed and elucidated.

The industrial process for the production of steviol glycosides is often the least understood aspect of the entire value-added chain of stevia-derived products. All the information available about the production process is sometimes too technical for a general audience and infested with false affirmation from uninformed persons or by lobbying actions.

3.2 Steviol Glycosides Production Technologies

Many different procedures for the extraction and purification of steviol glycosides from dried *Stevia rebaudiana* leaves in various degrees of purification have been published in the literature and there is still a lot of work to be done for new separation or purification technologies of minor steviol glycosides. The majority of these technologies has been described in the patent literature starting in Japan in the 1970s[64,72] and has increased in the years 1990–2011.[30,46,47,69] More recently, the level of patents and publications has started to increase again with the appearance of new biotechnological techniques for steviol glycoside production.[55–60,79,86,111] However, today most of the manufacturers still use what could be called "the conventional or traditional production process".[40] This production is also laid down in several regulatory documents such as U.S. Food & Drug Administration (FDA) GRAS notices.[94–98] They all refer to the description given at the 73rd JECFA meeting in 2010[87] but where the process is only roughly described. The production process involves several steps that are described in the following paragraphs.

3.2.1 Objectives

To understand the reason for the use of the different processes it is important to take into account the aims. For a food additive like steviol glycosides, the food industry must achieve the following objectives:

- obtaining a product with the best sweetening power with a good acceptable taste as close as possible to sugar
- optimising the process regarding yield and extraction time
- obtaining a product that is natural, which means that the components of the final product have to be unmodified all along the process lines; no chemical reaction or transformation of the components may occur
- achieving all of that at a reasonable price for the customers/users and with a good profit for the producer and all the participants in the supply chain.

The technology used for steviol glycosides extraction and purification must be chosen in order to reach these objectives.

3.2.2 Process Steps

For the extraction/purification of steviol glycosides the objective is to separate them from the rest of the plant constituents. As steviol glycosides are soluble in polar solvents, this process is a solid–liquid extraction purification. Today, this process can be roughly separated in three major steps: (1) primary extraction; (2) separation and (3) final purification of steviol glycosides.

Depending on the reachable purity, the procedure is significantly different with more or less technical difficulties. The classical process is used by most of the manufacturers at the time being; this process can be described as following (Figure 3.1).

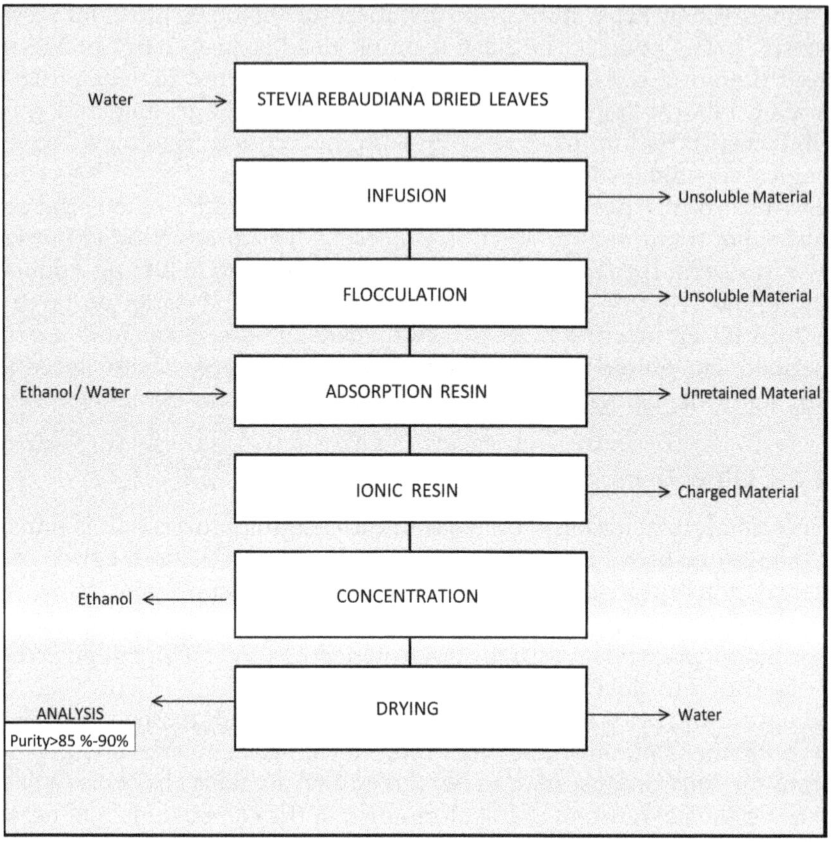

Figure 3.1 Simplified flow chart process for primary extraction and separation of steviol glycosides from *Stevia rebaudiana* dried leaves.

3.2.3 Steviol Glycosides Primary Extraction

Primary extraction is a solid–liquid extraction with the plant itself (or parts of it) as solid material and an appropriate solvent as the liquid part. As steviol glycosides have good solubility in polar solvents like pure water or ethanol (or mixtures of them), these are used. Primary extraction can be done from fresh[44] or from mostly dried material, from the whole harvested plant or from only separated leaves. The yield, purity, and costs will be different depending on the starting material. As steviol glycoside concentration is highest in the leaves (leaf vacuoles),[29,41] the optimised process uses isolated leaves instead of the complete plant (or other parts of it like the stems and petioles).

Leaves pretreatment: To optimise the extraction process, dried stevia leaves are used: as whole leaves, crushed, fine cut or fine powdered after grinding. Often a fine powder of particle size 0.5–1 mm size is used[25,50,92,112,114] allowing for good accessibility and diffusion of the steviol glycosides without lumps.

In our experience, a particle size of 1 mm resulted in a yield of 95.3%, which could be increased to 100.8% by reducing the size under 0.5 mm.[37] Dried leaves could be pre-treated with organic non-polar solvents like chloroform, ethylene chloride or hexane to remove fats and waxes from the leaves before steviol glycosides extraction.[72]

Steviol glycosides extraction: Most of the recent technologies start with an infusion as primary extraction. The operation consists of an extraction of dried leaves (pre-treated or not), with a polar solvent. The most commonly used polar solvents are water, alcohols (ethanol or methanol) and mixtures of them.

For this first step a lot of different parameters have been tested in the laboratory or in pilot studies, such as plant–solvent ratio, solvent type,[4] infusion temperature, and infusion time. The yield of extraction must be determined according to these different parameters. The yield of extraction, a parameter that is often not mentioned in publications, is defined as the ratio between the total amount extracted and the total amount present in the leaves before extraction.[38] Most of the reports in the literature refer to laboratory experiments using optimal conditions where it is possible to extract nearly all of the steviol glycosides but these conditions imply the use of specific laboratory equipment or large quantities of solvent, which are not compatible with the economy (or ecology) of an industrial process. For example: under the following conditions: soaking the dried leaves into water at 90 °C for an hour with a ratio of solvent to leaves of 20 to 40 mL g^{-1} (v/w) or even more (10 L g^{-1}) and then washing the residue twice in the same way, the diffusion and concentration equilibrium allows for the extraction of almost all of the steviol glycosides from the leaves.[70] Parameters that influence the extraction yield have been tested at laboratory levels using pressurised liquid extraction (PLE) technology.[23,37,80,83] Temperature, time, plant–solvent ratio and number of extraction cycles are critical parameters for optimising the extraction yield. Table 3.1 shows some of the multiple operating conditions found in the literature. Particle size (leaves), type of solvent, temperature, extraction time and the liquid–solid ratio are some of the conditions that varied considerably depending on the study.

However, at the industrial level, the choice has to be made between the maximum yield and the cost and time of the process. This is why our industrial process uses simple conditions, water as the solvent (the cheapest and the safest), a temperature of 60–70 °C for energy savings, open tank batch infusion for minimal investment and an infusion time of one hour, which gives the optimal yield under such conditions. The batch process can be described as follows: the leaves are placed in the tank and hot or warm solvent is added to submerge the leaves. The system is then agitated and after a certain time, the extract is drained from the tank and the extracted leaves are discharged and pressed to recover the maximum amount of solvent containing steviol glycosides. A second and even third water extraction can be done to optimise the yield.

Table 3.1 Steviol glycoside extraction from *Stevia rebaudiana* using different methods (according to Jentzer[38])

Authors	Date	Temp. (°C)	Infusion time	Solvent–liquid ratio	Particle size (μm)	Solvent
Liu *et al.*	1997	64	7 h	ND[a]	ND	Methanol
Pasquel *et al.*	2000	25–100	1 h	20	ND	Water
Erkucuk *et al.*	2009	78–100	2 cycles (Soxhlet)	10	500	Water–methanol
Jaitak *et al.*	2009	Ambient	12 h	100	250	Methanol–water 80: 20
Teo *et al.*	2009	100	1 h	300	500	Water
Liu *et al.*	2010	100	2 h	10	1000	Water
Rai *et al.*	2012	78	56 min	14	ND	Water
Lorenzo *et al.*	2014	100	30 min	42	Not grinded	Water
Das *et al.*	2015	71	51 min	40	ND	Water
Periche *et al.*	2015	50	5 min	100	ND	Water

[a]ND = not communicated.

The leaves are then squeezed out either by basket centrifuges or screw presses. The extracted and spent leaves then go for solid waste disposal. The so-called tea is pooled with the drained liquid from the extractor tank.

For the batch process, further machinery could be used: jacketed closed tanks, vertical or rotating vessels or percolation vessels.

Alternatively, continuous systems have also been used like continuous counter current extractors: Horizontal-, Hildebrandt-, Bonotto-, Bollman- or Rotocell-type devices.[73]

All these systems have advantages and disadvantages, the continuous systems allow the process to be integrated into a semi-continuous process, but the extraction efficiency and the solvent consumption is not better than with batch systems. Finally, the cost of investment and the return on this investment will define the final choice.

3.2.3.1 Alternative Primary Extraction Technologies

In order to improve the extraction efficiency, solvent use, energy (or any other parameters that could help to increase the yield and reduce costs and ecologic impacts), alternative technologies have been tested. Examples include: pressurised hot water extraction (PHWE), ultrasound, microwaves, organic solvents, chromatography and enzyme-assisted extractions, supercritical fluid extraction (SFE), extrusion, high voltage electrical discharges and pulsed electric fields.[8] Some examples are presented in Table 3.2 according to Puri.[78]

Koubaa gives an extended overview about these different techniques, their principle and mechanism of work.[45] All of these techniques have been tested in the laboratory or at pilot plant level and even if they have shown good results in terms of purity and efficiency, no scale-up at an industrial

Table 3.2 Stevia leaves extraction operating conditions from different studies (according to Puri[78])

Method	Yield	Treatment time	Reference
SFE: supercritical fluid extraction			
SFE (CO_2–methanol = 80 : 20)	2.51 mg g^{-1}	NC	Choi *et al.* 2002
SFE (CO_2–methanol, water = 80 : 16 : 4)	3.56 mg g^{-1}	NC	
Organic solvent	2.36		
Chromatographic separation (hexane, dichloroethane, ethyl acetate, methanol)	3.53 mg per 600 g	24 h	Pol *et al.* 2007
PFE: pressurised fluid extraction			
Methanol	5.20%	50 min	Pol *et al.* 2007
Water	4.70%		
PHWE: pressurised hot water extraction	13.9 mg g^{-1}	50 min	
MAE: Microwave-assisted extraction	21.37 mg g^{-1}	20 min	Teo *et al.* 2009
Microwaves	8.64%	1 min	
Ultrasounds	4.2%	30 min	Jaitak *et al.* 2009
SFE: supercritical fluid extraction	36.66 mg g^{-1}	100 min	Erkucuk *et al.* 2009
Ethanol	48.6		
UAE: ultrasonic-assisted extraction	43.62%	32 min	Liu *et al.* 2010
HSCCC: high speed counter-current chromatography	26 µg per mg	270 min	Huang *et al.* 2010
Enzyme (hemicellulase)	369 µg per mg	45 min	Puri *et al.* 2011
Combination of three enzymes (hemicellulase, pectinase, cellulase)	975 µg per mg	76 min	

or pre-industrial level has been tested to our knowledge. Another difficulty lies in the costs associated with these technologies. The use of enzymes is very expensive for large quantities and the level of investment is very high for technologies like SFE, extrusion or even ultrasound- or microwave-assisted extractions. When dealing with large quantities, a lot of specialised equipment is needed.

Ultrasound-assisted extraction: Zlabur described a lot of advantages for the use of ultrasounds (between 20 and 100 kHz): flexibility in equipment, easy handling and cost effectiveness in combination with shorter treatment times, reduction of temperature, high reproducibility, improvement in mass transfer and an eco-friendly approach.[116] The reason for that is because ultrasound technology helps the diffusion of soluble molecules outside of the solid matrix by creating important energy levels inside the plant cells like cavitation and gas bubbles destroying the cell structure.[88] This technology helps the liberation of internal compounds into the solvent increasing the yield of extraction and reducing extraction time and applied energy.[8,31,70,108] Specially for steviol glycosides, ultrasounds allows for a reduction in time and extraction temperature giving a better yield in stevioside and rebaudioside A,[50,116] the same yield[5] or a lower yield for stevioside but a better one for rebaudioside A as compared to traditional extraction. The application of this technology could allow extraction yields higher than 90%.[90] The extraction can be run in a batch or a continuous probe system or in a sonification bath.[45,116]

Microwave-assisted extraction: microwaves (between 300 MHz and 300 GHz) also facilitate the diffusion of soluble molecules out of the solid matrix by generating an increase in temperature inside the plant material. The conduction and movement of molecules induced by the moving magnetic field induce molecular collisions that generate overheating. The heating of water and other molecules causes a swelling of the plant cells and plant cell wall disruption, thus releasing soluble molecules into the solvent. Like ultrasounds, microwaves increase the extraction yield and reduce the extraction time, energy and solvent.[5,8,36,90] Stevioside and rebaudioside A extraction can be improved by the use of microwaves: a better yield and shorter time compared to conventional extraction or ultrasounds use.[6,36,69,116] Jaitak[36] compared conventional (25 °C, 12 h) and ultrasound techniques (35 ± 5 °C, 30 min) with microwave treatment and achieved better yields for stevioside and rebaudioside A at 50 °C and 80 W for one minute. Periche[70] confirmed these findings and could even increase the yields by applying a power of 3.3 W g^{-1} for two minutes extraction time.

Supercritical fluid extraction (SFE): SFE has also been used for steviol glycoside extraction.[20,25] The supercritical state corresponds to the state of a compound at a temperature higher than its critical temperature and at a pressure higher than its critical pressure. In that stage liquid and gas are not discernible and the compound can diffuse easily throughout solid materials increasing diffusion and thus reducing extraction time. Carbon dioxide (CO_2) is the most commonly used fluid compound for that technology but it is not very polar and must be combined with co-solvents like water or alcohols to extract polar compounds like steviol glycosides. This can lead to better extraction yields when compared to conventional extraction methods.[20,25,68,114,116] SFE could present some advantages in terms of selectivity, extraction time and substitution of toxic organic solvents.[32] This extraction method could also be used prior to steviol glycosides extraction: supercritical CO_2 alone is used in a first step to extract non-polar components from the leaves followed by a second step using water as co-solvent to CO_2, thus helping a subsequent extraction of polar steviol glycosides. In that process, an enrichment of rebaudioside A compared to stevioside and minor steviol glycosides is observed resulting in a better taste compared to traditionally extracted stevia leaves.[68]

Subcritical water extraction: this process uses water under high pressure,[66] also called pressurised fluid extraction or pressurised hot water extraction (PHWE). The high pressure allows to keep the solvent in the liquid state above its boiling point. The subcritical stage of water lowers its polarity allowing the extraction of organic components; in that situation water acts as an organic solvent like ethanol or methanol.[33,93] PHWE is called a "green extraction" as it uses only water. Normally, no clean-up or other preparations are needed. The samples *e.g.* dried stevia leaves are put in the extractor and the extracted compounds can be used immediately for further treatments or preparations.[45] Steviol glycoside extraction using only water or a mixture of methanol–water has been reported to be optimal at a temperature of 110 °C under a pressure of 50 bars for 10 min.[74] The use of higher temperatures like 120 °C could degrade steviol glycosides.[93] Therefore, Jentzer[37] tested 100 °C as a highest temperature and could yield more than 95% steviol glycosides within 4 min. Other experiments confirmed that temperature is the most important parameter for yield optimisation.[37,93,113]

Pulsed electric technology assisted extraction: Two different techniques are available: pulsed electric fields (PEF) and high voltage electric discharges (HVED). Both techniques are intensively described by Koubaa and Barba and applied to steviol glycoside extraction with high yields.[9,45] These techniques, especially PEF, have already been scaled-up for application in numerous food industries.[9,65]

Twin-screw extrusion: Extrusion has also been tested using a twin-screw extruder (Figure 3.2). This technology has been used at the INP Toulouse, to extract essential oils, pigments, proteins or hemicelluloses from plant materials.[27,38] The twin-screw extruder helps in solvent penetration and cell wall disruption. Results obtained for steviol glycosides extraction using this technology show similar results as those obtained with the simple infusion but with the advantage of a continuous process. However, the investment cost for this equipment limits its application in industry.

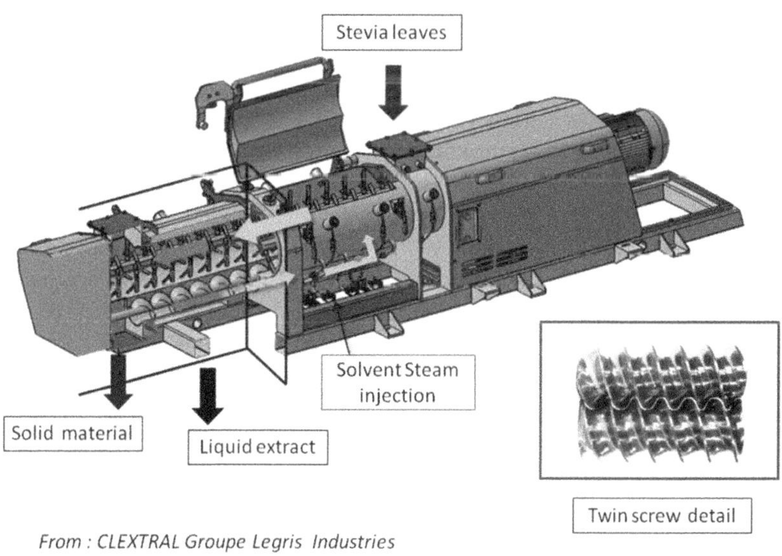

Figure 3.2 Schematic of the twin-screw extrusion principle.

Enzyme-assisted extraction: To improve the accessibility to the steviol glycosides located in the vacuoles of the plant leaves cells, enzymes that degrade the cell walls can be used. Under these conditions intracellular components become more accessible to solvents improving the extraction yield. Enzyme-assisted extraction is common in juice extraction from fruits. Cellulases, pectinases or hemicellulases are the most commonly used enzymes alone or in mixtures.[2,77,78] Steviol glycosides extraction with a mixture of cellulase, pectinase and hemicellulase (2% each) results in a better yield than with conventional methods and is even three times higher when compared to hemicellulase alone.[77]

3.2.3.2 Discussion

The aim of the primary extraction is to extract steviol glycosides with maximum yield and efficiency. Among the different technologies the efficiency may vary from 60% to more than 95% of the total amount of steviol glycosides from the *Stevia rebaudiana* leaves. Users face a difficult choice that has to be made depending on the final objective to reach; several parameters have to be taken into account concerning the composition of the final "crude primary extract".

Extraction yield: It is easy to understand that if you increase the extraction yield, you will also increase the amount of "unwanted molecules" in the extract that have to be eliminated in the subsequent phases of separation and purification of the steviol glycosides. A good equilibrium between quantity and purity is a parameter that has to be considered. A good extraction efficiency with a low purity (a lot of non-steviol glycoside molecules present in the extract) is not necessarily the best situation as this will impact negatively on the following processing phases: separation and purification.

Selectivity of the extraction: this is another parameter that could influence the choice of technology. Depending on the applied process, some steviol glycosides may be enriched in the extract (*e.g.* rebaudioside A *vs.* stevioside or minor steviol glycosides) which can have an impact on the organoleptic quality of the end product (Table 3.3). Moreover, unwanted molecules like labdanes or sterebins (yellow oil) could have a negative impact on the organoleptic characteristics of the end product.[41]

Costs of investment: this is another parameter that impacts the choice of the technology in the industrial production. The global production economy is largely impacted by the investment costs for some specific technologies like supercritical fluid extraction, ultrasounds, microwaves or twin-screw extrusion where specific machinery is needed with high investment costs. Return on investment (ROI) is an important parameter for the choice of the steviol glycosides production technology.

The costs of the process: The costs of the technological or processing aids (*e.g.* solvents and other chemicals) have a non-negligible impact on the final production costs of the end product, *e.g.* the purified steviol glycosides;

Table 3.3 Sweetening power and taste quality according to composition of steviol glycosides.

Steviol glycoside	Glucose residues[a] R1	Glucose residues R2	Sweetening power	Taste quality
Stevioside	1 glu	2 glu	150–300	0
Dulcoside A	1 glu	1 glu 1 rha	30–40	−2
Rebaudioside A	1 glu	3 glu	250–450	2
Rebaudioside C	1 glu	2 glu 1 rha	50–70	−1
Rebaudioside D	2 glu	3 glu	200–250	3
Rebaudioside F	2 glu	2 glu 1 xyl	ND[b]	2
Rebaudioside E	2 glu	2 glu	150–200	1

[a]Glu = glucose; rha = rhamnose; xyl = xylose. [b]ND: not determined.

furthermore, working time or work load generated by the process is also a question to address as the associated costs will increase the price of the final product.

Food safety: the use of some chemical solvents that could remain in the end product, even at trace levels could prevent the use of steviol glycosides as food ingredients. Some chemicals or processing aids used in the process may also induce chemical reactions and modifications of the product if they could not be mastered.

Environmental impact of the process: the use of very large quantities of solvents that improve yield generates a lot of effluents that have to be treated or discarded in an environmentally friendly way. The use of solvents like alcohols, further solvents or certain processing aids will also produce waste that has to be removed before release, generating further costs and negative environmental impact.

Finally, in our experience, we consider that the optimal technology for the industrial extraction of steviol glycosides from *Stevia rebaudiana* with a reasonable ratio of yield/cost/ROI is the infusion of dried leaves with hot water in a stirred batch or preferably in a continuous extraction system. Until now, most of the large-scale primary extraction processes being used in different extraction facilities are batch processes.

We have been working in the process development of this extraction stage for a long time and developed a counter current continuous solid–liquid extraction process with a non-sophisticated continuous extractor. This process is very efficient in energy and the extract concentration and composition are optimal for further steps leading to the end product. In this process, a large quantity of leaves can be extracted with a reasonable amount of water that could be used later for watering stevia cultures in the fields. With this technology and a reasonable liquid–solid ratio (10 : 1, w/w) we are able to extract 85–90% of the total steviol glycosides present in the dried *Stevia rebaudiana* leaves in an economical way.[71]

3.2.4 Steviol Glycosides Separation

Primary extraction provides a crude, dark brown extract containing steviol glycosides and other polar soluble constituents like carbohydrates, polyphenols, proteins, fibres, dyes, salts *etc.* The next step of the process aims to separate steviol glycosides from these impurities extracted at the same time during the primary extraction. This part of the process is sometimes named 'clarification' as it leads to a clear extract. A simple clarification method consists of centrifugation and microfiltration,[19] however, the efficiency of this methodology is low and more operations have to be done afterwards.

3.2.4.1 The Classical Process

The classical process is done in two steps: (1) a flocculation and then (2) a resin chromatography (Figure 3.1).

Flocculation: First, macromolecules like carbohydrates, proteins, plant pigments are precipitated using flocculants like calcium hydroxide (up to pH 10), aluminium hydroxide, calcium carbonate, ferric chloride or other chelating agents, allowing these molecules to precipitate. The precipitate is then filtrated or centrifuged, and the pH adjusted to neutrality before starting the chromatographic phase. Alternatively, an electro coagulation process could be used for this clarification step.[15,16,39]

Separation resins: After filtration or flocculation, the resulting tea is poured on a series of chromatographic columns with resins (adsorption, ionic)[7,13,28,94,95] or into a resin exchange tank.[99] The type and order of resins depends on the employed process. The principle is to adsorb steviol glycosides (polar molecules) on a polar resin and after the unwanted components have run through by washing with water, to desorb them using ethanol–water (50–65 : 50–35; v/v). Steviol glycosides adsorbed on polar resins are separated from other soluble molecules depending on their polarity. Ionic resins remove charged molecules like salts and pigments and thus complete the separation of steviol glycosides from other plant molecules. Different ionic resins can be used, strong anionic or cationic, weak anionic or cationic resins or mixed beds in order to complete separation of steviol glycosides from unwanted molecules.

Different types of polar and ionic resins are used in that process and can be purchased from different suppliers with efficiency that can differ depending on the functional groups used in these resins.[18] Mixed-mode microporous adsorption resins (MARs) have been used to increase the efficiency of separation and to enrich the eluate in rebaudioside A content.[18,51] Modified zeolites have also been used to clarify *Stevia rebaudiana* crude extracts.[53,63]

The application of active charcoal is also described in order to remove unwanted molecules. However, in our experience such a step has a negative impact on the yield as the active charcoal we have used also retains a significant amount of steviol glycosides.[71] Many different active charcoals exist with different porosity, origin or reactivity that makes the right choice quite difficult.

Recently, systems with several columns with different type of resins have been used to enrich eluates with specific steviol glycosides like rebaudioside D or rebaudioside M.[55]

At the end of this separation step, the extract containing the purified steviol glycosides in an alcohol–water solution is concentrated and then dried by evaporation and/or spray drying, the powder obtained contains steviol glycosides at a purity of 85–90%.

3.2.4.1.1 Solvent Separation. In that alternative process, the flocculated clarified extract is separated in a continuous counter current liquid–liquid extraction column using water and an immiscible organic solvent. The used solvent has a higher affinity towards steviol glycosides than water, so that steviol glycosides migrate from the water in the organic solvent. After separation of the water–solvent phases, steviol glycosides are recovered in the solvent fraction, which is then treated with active charcoal to remove pigments and other impurities. After evaporation of the solvent under vacuum, steviol glycosides crystallise in that described process. To recover steviol glycosides, a liquid–solid separation is carried out in a basket centrifuge and then crystallised steviol glycosides are dried by heating under vacuum.[85]

3.2.4.1.2 Membrane Filtration Technology. Some promising experiments have been done using membrane filtration to achieve steviol glycosides separation and even complete purification.[7,13,84] Here, the steviol glycoside molecules are separated according to their molecular weight and size using filtration, microfiltration, ultra-filtration, fine ultra-filtration (or nanofiltration) and reverse osmosis. Figure 3.3 shows the principle of that process.

As we know well, the composition and the size of the molecules to be purified are important to know. It is theoretically possible to choose a battery of membranes (micro filtration, ultra-filtration, nanofiltration, reverse osmosis) in order to separate steviol glycosides molecules from other plant materials, minerals or other small molecules. However, the variety of the filtration materials as organic or mineral membranes, shape and structures of these membranes makes the choice difficult and the comparison between results obtained in different places with different materials difficult to reproduce or even to interpret.[62,81,82] Another problem is the fouling of the membranes leading at least to a blocking of the membranes.[82]

Gonzalez gave an overview about studies related to the use of filtration membranes for the preparation of a purified stevia extract and even published a general scheme for the purification, which is based on the work of Zhang and Vanneste.[31,107,115]

As far as we know, these filtration technologies do not achieve the goal in terms of purity (at least for 95% purity) except in some laboratory cases.[81,83] The advantage of such a technology lies in the fact that it is theoretically possible to use only water all along the process of extraction separation and purification of steviol glycosides, establishing a "green" method for the production of steviol glycosides. It has also been claimed that some interesting

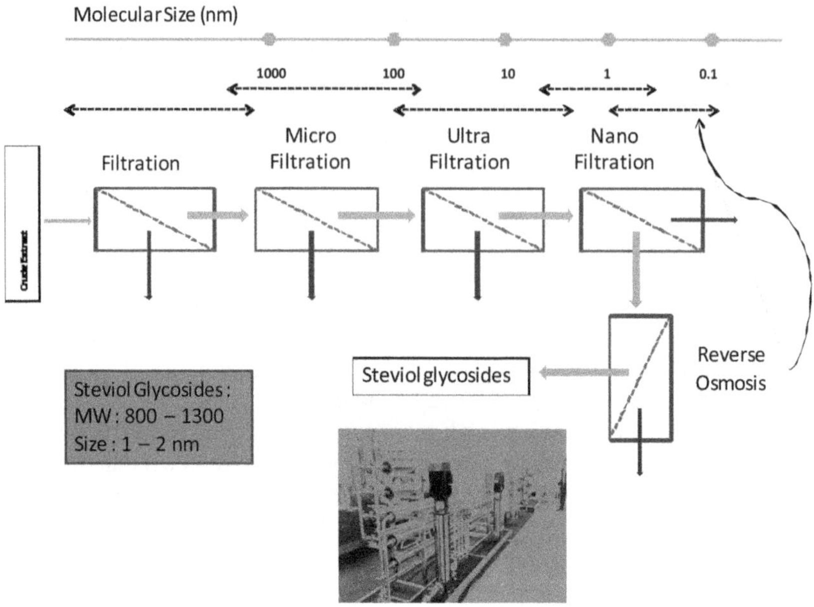

Figure 3.3 Membrane filtration principles.

molecules in terms of biological activities (antioxidants) could be retained in such a purification process.[91]

Some reports show that if the yield could be interesting (between 70 to 80%), the purity of the steviol glycosides remains low (between 10 to 35% *vs.* dry matter) which does not fulfil the requirements of the regulations in different countries.[7,44]

Another point is the taste of the purified extract. Comparing stevia extracts obtained by membrane purification (MP), enzyme modification and classical ethanol crystallisation, it could be shown that MP products had strong off-flavour attributes like bitterness, liquorice and metallic.[26]

3.2.4.1.3 Discussion. Whatever the process or the equipment used in this first part of production—extraction and separation of the steviol glycosides—the final result is a powder between 80 and 90% purity. Many claims have been made saying that 95% purity on dry matter (that is the purity required in *e.g.* European regulation) can be reached, but as far as we know these experiments have only been done in the laboratory under very specific conditions or on a pilot plant scale with results and yields incompatible with an industrial process that is economically viable.

3.2.5 Steviol Glycosides Final Purification

Starting from a dry steviol glycosides preparation with a purity between 80 and 90% in dry matter, a purification process has to be used in order to reach a purity *vs.* dry matter equal or higher than 95% (for regulatory

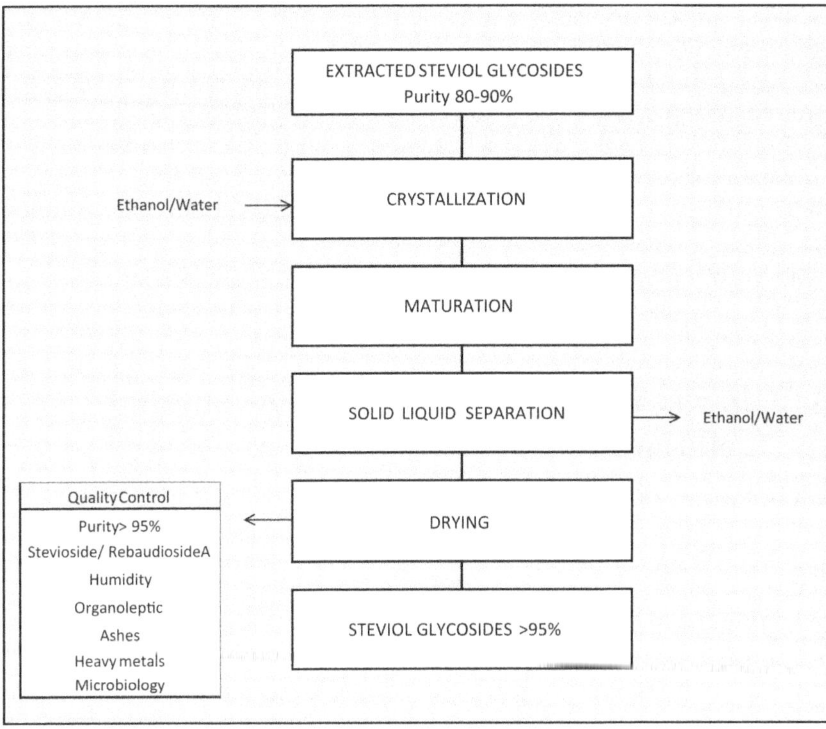

Figure 3.4 Simplified flow chart process for the purification of steviol glycosides to 95% purity.

reasons). This part of the process is mainly done by crystallisation in alcohols.[52] The crystallisation of heterosides in alcohol was first described for steviol glycosides by Bridel and Laviel in 1931.[12] Since then, many different protocols have been patented or published describing processes to purify steviol glycosides all together or separately like *e.g.* rebaudioside A. A saturated solution of the primary extract (steviol glycosides 80–90% purity) in ethanol, a mix of alcohols and water is heated in a reflux system. During this process crystallisation starts, and the solution becomes "milky". A maturation phase with a time-controlled decrease in temperature is then conducted. A simplified flow chart of this process is presented in Figure 3.4.[71]

Depending on the parameters: alcohol type, alcohol mixtures, water content, temperature, time and ratio of stevia extract to solvent, the obtained results differ, and it is possible to selectively separate some steviol glycosides from the others (rebaudioside A alone, stevioside, minor steviol glycosides) or receiving a pattern similar to that being originally present in the stevia leaves. The following examples can elucidate this issue: *e.g.* the best yield for rebaudioside A at a 95.6% purity level can be obtained with 88% ethanol and a ratio of stevia extract to solvent of 1 : 3 while rebaudioside D content increases with an increasing concentration of ethanol (up to 88–90%) and a decreasing ratio,[3] and using methanol instead of ethanol for crystallisation yields in purified rebaudioside M, which is said to be the best tasting steviol glycoside but being only present in the plant with an amount of ≤1%.[100] Therefore, it is possible to purify nearly every steviol glycoside alone or combinations of them to the desired purity conform to the required regulation, *i.e.* rebaudioside A with 97% purity *vs.* dry matter[95] or total steviol glycosides with 95% purity *vs.* dry matter (EU regulation[21,24] and some other countries since 2012).

Final purity is measured by a HPLC analytical method officially described by the JECFA[87] and improved for efficiency and accuracy.[49,61,110]

3.3 Alternative Steviol Glycosides Production Techniques

The production of a highly efficient sweetener with a high sweetening power from stevia leaves and decreasing (at the same time) unwanted organoleptic characteristics is a goal that many researchers and companies try to reach. As the sweetening power is mainly related to the number and type of sugar units linked to the steviol aglycone, the addition of glucose to this diterpene skeleton has now been realised by enzymatic modification.[89] Chemical modifications (hydrolysis) have also been used to enrich purified steviol glycosides in minor steviol glycosides with better organoleptic properties like rebaudioside B or steviol glycosides isomers.[17,47] Finding the right steviol glycoside profile is also a target for plant breeders who try to select *Stevia rebaudiana* varieties with different steviol glycosides patterns.[10] Table 3.4 gives an example for successful breeding of new varieties with an increased content of the better tasting steviol glycosides rebaudioside D and M.

Table 3.4 Increase of rebaudioside D and M content by improvement of breeding.

Variety	Stevioside[a]	Reb A[a]	Reb D[a]	Reb M[a]	Reference
S. rebaudiana (Criolla)	7.9%	4.9%			Wölwer-Rieck, 2010
S. rebaudiana (Morita)	Up to 4.1%	Up to 12.7%	Under 0.5%		Morita, US Patent 2000/6,050,561 Otha, 2010
S. rebaudiana (AKH L1)	1.3%	11.5%			US Patent 2012/ PP23,164 P3
S. rebaudiana	0.65–1.3%	4.6–15%	0.5–2.0%	0.4–0.6% sum of Reb E+N+M	Patent WO 2014/146084
S. rebaudiana			0.6–3.3%	0.5–1.2%	Patent WO 2016/049531
S. rebaudiana				>1.0%	Patent filling

[a]Percentage of single steviol glycosides of dry weight; with non-GMO technology.

3.3.1 Enzymatic Modification

CGTase: Cyclomaltodextrin-glucosyltransferase (β-CGTase) catalyses the glycosylation of steviol glycosides by transferring α-glucosyl units from starch to a glucosyl moiety at the 4-hydroxyl group (*trans*-α-1,4-D-glucosylation). This enzyme produced by several Bacillus species, *e.g. B. stearothermophilus*, *B. megaterium*, *B. macerans*, *B. circulans*, *B. halophilus* and *B. alcalophilus*, is used to add glucose moieties to steviol glycosides yielding a complex mixture with several glucose residues linked both to the 19-*O* glucosyl unit and to the 13-*O*-sophorosyl moiety.[1,43,79] The sugar source for glycosylation is maltodextrin coming from starch (tapioca, maize, wheat and others). The results of the enzymatic reaction are the so called glycosylated steviol glycosides (GSGs) that can have 6, 7, 8 and more glucose moieties linked to the steviol backbone. It could be shown that there is no linear relationship between the sweetening power and the number of glucose attached and that if the number of attached sugar units is too high the sweetness even decreases. The effect on bitterness of the steviol glycosides has also been demonstrated.[79] Mastering this reaction in order to obtain the desired GSGs composition is a challenge that researchers are still trying to achieve. However, these GSGs, which are not allowed as sweeteners in some countries (within the EU for example) but got their GRAS status in the US[101–103] could be used as flavouring substances. Furthermore, it has to be mentioned that this is a chemical conversion of the natural-based sweetener. The pattern and amount of originally occurring steviol glycosides have been changed and further new steviol glycosides not existing in the plant have been formed (Table 3.5).

UGT enzymes: UGT enzymes are essential parts of the biochemical pathway of steviol glycosides. They add glycosides units to the steviol aglycone

Table 3.5 Chemical changes in the steviol glycoside pattern by enzymatic modification (Part of Table 5 in GRAS notice No. 448[102]).

Component	Untreated steviol glycosides (%)	Range of glucosylated steviol glycosides (%)
Steviolbioside	3.6–5.6	2.2–3.2
Dulcoside A	0.7–1.3	0.2–0.4
Rebaudioside B	0.3–0.7	0.6–1.2
Stevioside	78.7–81.2	15.3–16.6
Rebaudioside C	1.6–2.4	0.6–1.0
Rebaudioside F	0.3–0.5	0.1–0.2
Rebaudioside A	10.0–11.8	3.3–5.0
Glucosylstevioside *m*/*z* 966	ND	17.1–18.0
Glucosylstevioside *m*/*z* 1112	ND	1.5–3.7
Glucosylstevioside *m*/*z* 1128	ND	16.6–17.2
Glucosylstevioside *m*/*z* 1274	ND	1.4–2.7
Glucosylstevioside *m*/*z* 1290	ND	14.5–17.8
Glucosylstevioside *m*/*z* 1452	ND	10.6–12.3
Glucosylstevioside *m*/*z* > 1452	ND	3.9–6.9

in the stevia plant cells to synthesise the different steviol glycosides.[11,29] These enzymes could be used to obtain steviol glycosides from steviol and glucosides donors either by using purified proteins or enzymes from *Stevia rebaudiana* (or other plants) or after production by genetically modified microorganisms that are able to express genes coding for these proteins.[48] For example, by this alternative technology (starting with purified stevia extract produced in the traditional way) rebaudioside M can be produced by using genetically modified baker's yeast. This technology received GRAS status in 2017.[104]

3.3.2 Genetic Engineering and Fermentation Techniques

More recently, the use of genetic engineering and fermentation has been published in order to produce better tasting steviol glycosides like rebaudioside D and M (that is the same as X).[75,76] The company Evolva in Switzerland, associated with Cargill, granted a European Patent: "Methods for improved production of rebaudioside D and rebaudioside M"[60] and some other international patent applications: "Recombinant Production of Steviol Glycosides",[35] "Steviol glycosides compositions sensory properties".[34] Stephanopoulos[86] also patented the "Microbial production of natural sweeteners, diterpenoid steviol glycosides" in 2016. Markosyan (Pure Circle company) patented a method for producing Rebaudioside X.[57]

These technologies are based on past and present knowledge of the biochemical pathway and enzymes for steviol glycosides biosynthesis.[11] The isolation of the involved genes from *Stevia rebaudiana* or other plants and the transfer and expression of these genes into a microorganism allow for the production of minor steviol glycosides by cultivation of the said microorganism, mainly rebaudioside D and rebaudioside M. Isolation and cloning of the genes involved in the upstream isoprenoid pathway,[48] a down streaming rebaudioside D pathway including the required UDP-glucosyl transferases (UGT74G, UGT76G, UGT85C, UGT91D and more) under the control of specific operons in microorganisms like bacteria: *Escherichia coli*, *Bacillus* spp.; yeasts: *Saccharomyces cerevisiae*, *Yarrowia lipolytica*; fungi: *Aspergillus* spp., *Penicillium* spp., *Pichia* spp.; as claimed by the authors, algae or plant cells may allow the production of steviol and steviol glycosides that could be recovered for the culture broth, or to produce UGT enzymes that are used to glycosylate purified steviol glycosides into highly glycosylated steviol glycosides.

These patents and patent applications are claiming the use of microorganisms but also plant cells and plants for the production of enzymes, steviol and steviol glycosides. However, in the present stage, only bacteria and yeast can produce UGT enzymes or steviol glycosides by fermentation. Two applications got the GRAS status in the US[105,106] using genetically modified baker's yeast and yeast from *Yarrowia lipolytica*. Both technologies renounce completely the plant and its leaves and are only produced *via* fermentation.

The biotechnological industrial production of such steviol glycosides and the sales of these better tasting steviol glycosides were announced for 2016, but it's only in March 2018 that Cargill and Evolva announced the launch of EverSweet.[109] Composed of rebaudioside M and rebaudioside D produced by fermentation using a genetically modified baker's yeast and corn dextrose as the substrate, they are now available in commercial quantities and will be sold and industrially produced in larger quantities by the end of the year. There are some reasons for this delay: first is the regulation that is not always clear in different countries for example in the US or in Europe, and second, it has to be demonstrated that the costs of these productions are competitive with those of the traditional production *via* cultivation of the plant, harvesting the leaves and extraction. On this question the debate is still open between fermentation and breeding of stevia varieties with higher content of rebaudioside M and D. Whatever the production of steviol glycosides in genetically engineered microorganisms or selected plants is efficient and even if these steviol glycosides are excreted in the culture medium, it will be still necessary to purify them from other unwanted molecules, thus coming back to extraction and purification technologies described in that chapter.

3.4 Production Technologies *vs.* Naturalness

For the consumers and, by consequence for the food industry, a 100% natural product is the goal to achieve for evident reasons. But the term "natural" is under discussion all over the world. No precise definition has been given. For steviol glycosides (as food additive) it is forbidden in Europe to label them as "natural sweeteners" due to the complex purification and extraction steps in which "Several other related steviol glycosides that may be generated as a result of the production process, but do not occur naturally in the *Stevia rebaudiana* plant have been identified in small amounts (0.10–0.37% w/w)".[21,22] The use of ionic resin during the process can lead to the formation of these molecules. Some people claim that steviol glycosides cannot be natural because of the occurrence of these isomers. However, in nature it is completely normal that small amounts of isomers occur. Nature is never perfect and, if a beta-isomer is formed in a biosynthetic pathway, the chances are realistic that small amounts of alpha-isomers are also formed. A lot of examples can be found in nature *e.g.* as flavour compounds that, as a rule, occur as alpha- and beta-isomers. Moreover, people claiming that the isomers in steviol glycosides are not natural, should give a positive proof for this because, scientifically, it could not be demonstrated till now that isomers are not present. They should prove that the amounts of isomers in a processed product are much higher than in the plant itself. A recent study has shown that the most abundant steviol glycosides as described by JECFA[87] and produced by the classical extraction and separation method are strictly identical to those present in the Stevia leaves reinforcing the natural character of these molecules.[67]

Another point is the fact that steviol glycosides as food additives cannot be certified as organic in Europe due to the resins used during the purification process. This is regarded differently in other countries: in the US and many other countries like China, purified steviol glycosides are considered as a natural product and may even be certified as organic.

Furthermore, it has been shown that some steviol glycosides isomers have a better taste profile then the original molecules, and so some companies try to produce these isomers.[47]

3.5 Conclusion

According to European regulation[21] the specifications for steviol glycosides includes a definition of the production process:

> *"The manufacturing process comprises two main phases: the first involving water extraction of the leaves of the* Stevia rebaudiana *Bertoni plants and preliminary purification of the extract by employing ion exchange chromatography to yield a steviol glycosides primary extract, and the second involving crystallisation of the steviol glycosides from methanol or aqueous ethanol resulting in a final product..."*

Obviously, this description corresponds to the "traditional" production process employed by most steviol glycoside producing companies. Does that mean that, at least in Europe, this is the only way to produce steviol glycosides from *Stevia rebaudiana* leaves and, if so, what is about steviol glycosides produced by other technologies described in this chapter, some of them already used by some companies in the world? When questioned, the authorities in Europe give this answer: "*these products will need a new authorisation by filing a new dossier*"; if this answer is not shared by all European member states, and elsewhere in the world, it is still unknown what will be the status of steviol glycosides produced by other processes like enzymatic modification or fermentation of genetically modified organisms or even genetically modified plants. And what is about the new EU regulation[22] considering that minor steviol glycosides like rebaudioside M existing in the plant can be added to the existing specifications knowing that in order to produce these minor steviol glycosides in significant quantities, a different process as the traditional one like biotechnology or genetically modified organisms has to be applied?

Still, looking at the definition of the production process in the European regulation we can read:

> *"The additive may content residues of ion exchange resins used in the manufacturing process. Several other related steviol glycosides may be generated as a result of the production process, but do not occur naturally in the stevia plant have been identified in small amounts (0.1–0.37% w/w)..."*

This means that the regulation considers that this process modifies the components present in the plant and that the product may contain residues of exchange resins. Not only the presence of such residues from exchange resins nor the generation throughout the process of several other "related steviol glycosides" have been scientifically demonstrated, it still has to be proven that these steviol glycosides isomers are not present naturally in the plant. Finally, if an EFSA panel considers that these minor steviol glycosides are not harmful for human consumption, how do we explain this restriction in the regulation?

A recent patent[15] describes an extraction method for providing an organic certifiable *Stevia rebaudiana* extract; this patent shows that depending on the perception of the impact of extraction technologies the finish product could or cannot be "organic"!

The GMO issue: This question is again treated differently in different countries around the world depending on the interpretation of local regulations. The presentation to the consumers throughout labelling of the products is subject of a lot of creativity on how to label the ingredient in a friendly way according to regulation: "specially crafted baker's yeast" instead of "genetically engineered baker's yeast" is an exemple.[109] The fact that, in the final product, the genetically modified organism, the plant or the microorganism, is completely removed from the purified ingredient implies, according to local GMO labelling legislation, that the GMO label is not needed; but under the Non-GMO Project ingredients produced by "synthetic biology" cannot qualify for this Non-GMO label. If the ingredient has been proven to be safe (which is still scientifically questionable) does that mean that the consumer does not need to know how it has been produced? This is no longer a scientific or technological question but a societal behavior: today more than before, science and technology must be understood and accepted by consumers.

These observations give rise to questions regarding steviol glycosides naturalness and organic certification. The world differences in treatment of this issue represent today an issue in consumers' understanding about steviol glycosides. Are they natural products or only a new synthetic sweetener like most of the other additives in the E900 family? A common definition of a natural product has to be agreed at a global level and this could be done simply, not with hundreds of regulations. A natural product is a product that exists in nature, it is not modified by the production process in a significant manner and it is not synthesised in any way from other substances. This is simple, easy to understand by everybody and easily controlled.

Steviol glycosides production technology definitely has a major impact on their naturalness and therefore on their uses by the consumers for their wellbeing.

Acknowledgements

The support from Dr Ursula Wölwer-Rieck for discussions and corrections of this manuscript is acknowledged.

References

1. V. H. Abeylyan, A. M. Balayan, V. T. Ghochikyan and A. A. Markosyan, *Appl. Biochem. Microbiol.*, 2004, **40**, 129.
2. V. H. Abeylyan, *et al.*, *US Pat.* 20060134292, 2006.
3. V. H. Abeylyan, *et al.*, *US Pat.* 20110092684, 2011.
4. A. Afandi, S. Sarijan and R. K. Shaha, *J. Trop. Resour. Sustainable Sci.*, 2013, **1**, 62.
5. A. Alupului, I. Calinescu and V. Larvic, *AIDC Conf. Ser.*, 2009, **9**, 1.
6. K. Ameer, S. W. Bae, Y. Jo, H. G. Lee, A. Ameer and J. H. Kwon, *Food Chem.*, 2017, **229**, 198.
7. M. Anvari and G. Khayati, *Pol. J. Chem. Technol.*, 2016, **18**, 127.
8. J. Azmir, I. S. M. Zaidul, M. M. Rahman, K. M. Sharif, A. Mohamed, F. Sahena, M. H. A. Jahurul, K. Ghafoor, N. A. N. Norulaini and A. K. M. Omar, *J. Food Eng.*, 2013, **117**, 426.
9. F. J. Barba, N. Grimi and E. Vorobier, *J. Food Eng.*, 2015, **149**, 222.
10. C. Barbet-Massin, S. Giuliano, L. Aletto, J. Dayde and M. Berger, *Genet. Resour. Crop Evol.*, 2016, **63**(4), 685.
11. J. E. Brandle and P. G. Telmer, *Phytochemistry*, 2007, **68**, 1855.
12. L. R. Bridel and M. Bridel, *C. R. Acad. Sci.*, 1931, **192**, 1123.
13. S. J. Catani, *US Pat.* 20130108718, 2013.
14. S. Ceunen and J. M. C. Geuns, *J. Nat. Prod.*, 2013, **76**(6), 1201.
15. S. Chabot, *et al.*, *US Pat.*, No 8 728 545, 2014.
16. N. Chairungsi, K. Jumpatong, W. Phutdhawong and D. Buddhasukh, *Molecules*, 2006, **11**, 309.
17. V. S. Chaturvedula and I. Prakash, *Int. J. Biol. Med. Res.*, 2011, **2**(4), 1090.
18. Z. Chen, X. Wei, J. Li and D. Di, *Sep. Purif. Technol.*, 2012, **89**, 22.
19. G. C. Chhaya, G. C. Majumdar and S. De, *Sep. Sci. Technol.*, 2013, **48**, 113.
20. Y. H. Choi, I. Kim, K. D. Yoon, S. J. Lee, C. Y. Kim, K. P. Yoo, Y. H Choi and J. Kim, *Chromatographia*, 2002, **55**(9–10), 617.
21. Commission regulation (EU) 231/2012, *Off. J. Eur. Union*, 2012, L 83/1–295.
22. Commission regulation (EU) 2016/1814, *Off. J. Eur. Union*, 2016, L 278/37.
23. A. Das, A. K. Golder and C. Das, *Ind. Crops Prod.*, 2015, **65**, 415.
24. EFSA Panel on Food Addiditives, Scientific Opinion, *EFSA J.*, 2015, **13**(12), 4316.
25. A. Erkucuk, I. H. Akgun and O. Yesil-Celiktas, *J. Supercrit. Fluids*, 2009, **51**, 29.
26. M. I. Espinoza, J. P. Vincken, M. Sanders, C. Castro, M. Stieger and E. Agosin, *J. Agric. Food Chem.*, 2014, **62**, 11797.
27. P. Evon, PhD thesis, INP Toulouse, No 2607, 2008.
28. P. Gandhi, *et al.*, *US Pat.* 20160015066, 2016.
29. J. M. Geuns, *Stevia and Steviol Glycosides*, Euprint, Heverlee, Belgium, 2010.
30. R. H. Giovanetto, *US Pat.*, No 4 892 938, 1990.

31. C. González, M. Tapia, E. Pérez, D. Pallet and M. Dornier, *Fruits*, 2014, **69**, 127.
32. A. Hartmann and M. Ganzera, *Planta Med.*, 2015, **81**, 1570.
33. M. Herrero, A. Cifuentes and E. Ibanez, *Food Chem.*, 2006, **98**(1), 136.
34. P. Hicks, *et al.*, *WO Pat.* 2014/086890, 2014.
35. J. Houghton-Larsen, *et al.*, *WO Pat.* 2013/022989, 2013.
36. V. Jaitak, B. S. Bandna and V. K. Kaul, *Phytochem. Anal.*, 2009, **20**, 240.
37. J. B. Jentzer, M. Alignan, C. Vaca-Garcia, L. Rigal and G. Vilarem, *Food Chem.*, 2015, **166**, 561.
38. J. B. Jentzer, PhD thesis, INP Toulouse, No 0115, 2015.
39. K. Jumpatong, W. Phutdhawong and D. Buddhasukh, *Molecules*, 2006, **11**, 156.
40. M. Kassing, U. Jenelten, J. Schenk and J. Strube, *Chem. Eng. Technol.*, 2010, **33**(3), 377.
41. A. D. Kinghorn, *The Genus Stevia*, ed. A. D. Kinghorn, Taylor & Francis, London, 2002, p. 1.
42. A. D. Kinghorn and D. D. Soejarto, *Econ. Med. Plant Res.*, 1985, **1**, 1.
43. V. T. Kochikyan, A. A. Markosyan, L. A. Abelyan, A. M. Balayan and V. A. Abelyan, *Appl. Biochem. Microbiol.*, 2006, **42**(1), 31.
44. A. M. J. Koostra, H. J. H. Elissen and S. Huurman, *ACRRES-Wageningen*, 2015, Report PPO-632.
45. M. Koubaa, E. Roselló-Soto and J. Sic Zlabur, *et al.*, *J. Agric. Food Chem.*, 2015, **63**, 6835.
46. O. Kutowy, *et al.*, *US Pat.* 5972120, 1999.
47. T. Lee, *US Pat.* 7 964 232, 2011.
48. M. Lehmann, *et al.*, *WO Pat.* 2013/110673, 2013.
49. C. Lorenzo, J. Serrano-Diaz, M. Plaza, C. Quintanilla and G. L. Alonzo, *Food Chem.*, 2014, **157**, 518.
50. J. Liu, J. Li and J. Tang, *Food Bioprod. Process.*, 2010, **88**(2–3), 215.
51. Y. Liu, D. Di, Q. Bai, J. Li, Z. Chen, S. Lou and H. Ye, *J. Agric. Food Chem.*, 2011, **59**, 9629.
52. M. Magomet, *et al.*, *US Pat.* 20070082103, 2007.
53. I. C. C. Mantovanelli, E. C. Ferretti, M. R. Simoes and C. Ferreira da Silva, *Braz. J. Chem. Eng.*, 2004, **21**(3), 449.
54. K. Marcinek and Z. Krejpcio, *Acta Sci. Pol. Technol. Aliment.*, 2015, **14**(2), 145.
55. A. Markosyan, *et al.*, *US Pat.* 20130071339, 2013.
56. A. Markosyan, *et al.*, *US Pat.* 20140066400, 2014.
57. A. Markosyan, *et al.*, *US Pat.* 20150208704, 2015.
58. A. Markosyan, *et al.*, *US Pat.* 20150021869, 2015.
59. A. Markosyan, *et al.*, *US Pat.* 20140227421, 2014.
60. M. D. Mikkelsen, *et al.*, *US Pat.* 20160186225, 2016.
61. M. Molina-Calle, F. Priego-Capote and M. D. Luque de Castro, *Food Chem.*, 15/3/2017, **219**, 329–338, Accepted Manuscript.
62. S. Mondal, C. Rai and S. De, *Food Bioprocess Technol.*, 2013, **6**, 931.
63. E. P. Moraes and N. R. C. F. Machado, *Acta Sci.*, 2001, **23**(6), 1375.

64. T. Morita, *et al.*, *US Pat.* 4 082 858, 1978.
65. M. Negm, N. Grimi and E. Vorobiev, *International Congress on Green Extraction of Natural Products*-(GENP), 2013, Avignon, France.
66. A. Nieto, F. Borrull, E. Pocurull and R. M. Marcé, *TrAC, Trends Anal. Chem.*, 2010, **29**(7), 752.
67. A. Oehme, M. Wüst and U. Wölwer-Rieck, *Int. J. Food Sci. Technol.*, 2017, **52**, 2156.
68. A. Pasquel, M. A. A. Meireles, M. O. M. Marques and A. J. Petenate, *Braz. J. Chem. Eng.*, 2000, **17**(3), 271.
69. J. D. Payzant, *et al.*, *US Pat.* 5 962 678, 1999.
70. A. Periche, M. L. Castello, A. Heredia and I. Escriche, *Plant Foods Hum. Nutr.*, 2015, **70**(2), 119.
71. J. Perret, *et al.* unpublished results.
72. G. J. Persinos, *US Pat.* 3723410, 1973.
73. R. Poirot, PhD thesis, INP Toulouse, No 2570, 2007.
74. J. Pol, E. V. Ostra, P. Karasek, M. Roth, K. Benesova, P. Kotlarikova and J. Caslavsky, *Anal. Bioanal. Chem.*, 2007, **388**, 1847.
75. I. Prakash, C. Bunders, K. P. Devkota, R. D. Charan, C. Ramirez, C. Priedemann and A. Markosyan, *Biomolecules*, 2014, **4**, 374.
76. I. Prakash, A. Markosyan and C. Bunders, *Foods*, 2014, **3**, 162.
77. M. Puri, D. Sharma, C. J. Barrow and A. K. Tiwary, *Food Chem.*, 2012, **132**, 1113.
78. M. Puri, D. Sharma and A. K. Tiwari, *Biotechnol. Adv.*, 2011, **29**, 781.
79. S. Purkayastha, *et al.*, *US Pat.* 20150305379, 2015.
80. C. Rai, G. C. Majumbar and S. De, *Sep. Sci. Technol.*, 2012, **47**(7), 1014.
81. A. K. Rao, E. Prasad, G. Roopa, S. Sridhar and Y. V. L. Ravikumar, *Adv. Biosci. Biotechnol.*, 2012, **3**, 327.
82. M. H. M. Reis, F. V. Da Silva, C. M. G. Andrade, S. L. Rezende, M. R. Wolf Maciel and R. Bergamasco, *J. Food Process Eng.*, 2009, **32**, 338.
83. B. E. Richter, B. A. Jones, J. L. Ezzell, N. L. Porter, N. Avdalovic and C. Pohl, *Anal. Chem.*, 1996, **68**, 1033.
84. A. Roy, S. Moulik, S. Sridhar and S. De, *J. Food Sci. Technol.*, 2015, **52**(11), 7081.
85. C. Shantanu, Unpublished results, 2015, see https://www.steviashantanu.com/resin-free-stevia-extraction-technology (last accessed January 2018).
86. G. Stephanopoulos, *et al.*, *US Pat.* 20120164678, 2012.
87. Joint FAO/WHO Expert Committee on Food Additives (JECFA), 73rd Meeting, *Compendium of Food Additive Specifications: Steviol Glycosides*, Rome, FAO JECFA, Monographs, 2010, **10**, 17–22.
88. J. Sun, R. Gu and X. Yang, *WO Pat.* 2011/113373, 2011.
89. O. Tanaka, *Pure Appl. Chem.*, 1997, **69**(4), 675.
90. P. Tatke and Y. Jaiswal, *Res. J. Med. Plant*, 2011, **5**(1), 21.
91. S. Tavarini and L. G. Angelini, *J. Sci. Food Agric.*, 2013, **93**(9), 2121.
92. C. C. Teo, S. N. Tan, J. W. H. Yong, C. S. Hew and E. S. Ong, *J. Sep. Sci.*, 2009, 613.

93. C. C. Teo, S. N. Tan, J. W. H. Yong, C. S. Hew and E. S. Ong, *J. Chromatogr.*, 2010, **1217**(16), 2484.
94. U.S. Food & Drug Administration (FDA), GRAS notice No. 252, 2008.
95. U.S. Food & Drug Administration (FDA), GRAS notice No. 253, 2008.
96. U.S. Food & Drug Administration (FDA), GRAS notice No. 323, 2010.
97. U.S. Food & Drug Administration (FDA), GRAS notice No. 456, 2013.
98. U.S. Food & Drug Administration (FDA), GRAS notice No. 619, 2016.
99. U.S. Food & Drug Administration (FDA), GRAS notice No. 329, 2010.
100. U.S. Food & Drug Administration (FDA), GRAS notice No. 473, 2014.
101. U.S. Food & Drug Administration (FDA), GRAS notice No. 375, 2011.
102. U.S. Food & Drug Administration (FDA), GRAS notice No. 448, 2013.
103. U.S. Food & Drug Administration (FDA), GRAS notice No. 452, 2014.
104. U.S. Food & Drug Administration (FDA), GRAS notice No. 667, 2017.
105. U.S. Food & Drug Administration (FDA), GRAS notice No. 626, 2017.
106. U.S. Food & Drug Administration (FDA), GRAS notice No. 632, 2017.
107. J. Vanneste, A. Sotto and C. Courtin, *J. Food Eng.*, 2001, **103**, 285.
108. K. Vilkhu, R. Mawson, L. Simons and D. Bates, *Innovative Food Sci. Emerging Technol.*, 2008, **9**, 161.
109. E. Watson, Cargill launched EverSweet fermented steviol glycosides, see https://foodnavigator-usa.com/Article/2018/03/20/Cargill-launches-EverSweet-fermented-steviolglycosides (last accessed March 2018).
110. U. Wölwer-Rieck, *J. Agric. Food Chem.*, 2012, **60**(4), 886.
111. M. Yang, J. Hua and L. Qin, *US Pat.* 20080300402, 2008.
112. E. Yildiz-Ozturk, A. Nalbantsoy, O. Tag and O. Yesil-Celiktas, *Ind. Crops Prod.*, 2015, **77**, 961.
113. E. Yildiz-Ozturk, O. Tag and O. Yesil-Celiktas, *J. Supercrit. Fluids*, 2014, **95**, 422.
114. S. K. Yoda, M. O. Marques, A. J. Petenate and M. A. A. Meireles, *J. Food Eng.*, 2003, **57**(2), 125.
115. S. Zhang, S. Kumar and O. Kutowy, *Food Res. Int.*, 2000, **33**, 617.
116. J. S. Zlabur, S. Voca, N. Dobricevic, M. Brncic, F. Dujmic and S. R. Brncic, *Int. Agrophys.*, 2015, **29**, 231.

CHAPTER 4

Analysis of Steviol Glycosides

U. WÖLWER-RIECK*

Rheinische Friedrich-Wilhelms Universität Bonn, IEL, Bioanalytics/Food Chemistry, Endenicher Allee 11-13, 53115 Bonn, Germany
*E-mail: woelwer-rieck@uni-bonn.de

4.1 Introduction

Stevia rebaudiana accumulates more than 10 steviol glycosides in varying concentrations, depending on the genotype and cultivation conditions. Amounts of total steviol glycosides up to 20% of the dry leaf weight are reported.[1] The best known steviol glycosides are stevioside and rebaudioside A, which have the highest content in the plant (see Figure 4.1 and Table 4.1).

The analysis of steviol glycosides started in 1931 with the synthesis of stevioside described by Bridel and Lavieille elucidating the chemical formula of stevioside.[2] Some decades later, stevioside, rebaudioside A, B, C, D and E and dulcoside A could be identified.[3–7] Since then, many minor steviol glycosides have been detected. To date, more than 40 different steviol glycosides have been described, which are diterpenes based on the *ent*-kaur-16-en-19-oic acid skeleton and are bound to different sugar moieties at the C-13 and C-19 position (Figure 4.1).[8–10] They are divided into five groups, the glucosyl-, rhamnosyl-, xylosyl-, fructosyl- and deoxyglucose steviol family based on the type of glycosidic residues present. For example, stevioside and rebaudioside A belong to the glycosyl-family. Furthermore, several steviol glycosides have

Food Chemistry, Function and Analysis No. 7
Steviol Glycosides: Cultivation, Processing, Analysis and Applications in Food
Edited by Ursula Wölwer-Rieck

Published by the Royal Society of Chemistry, www.rsc.org

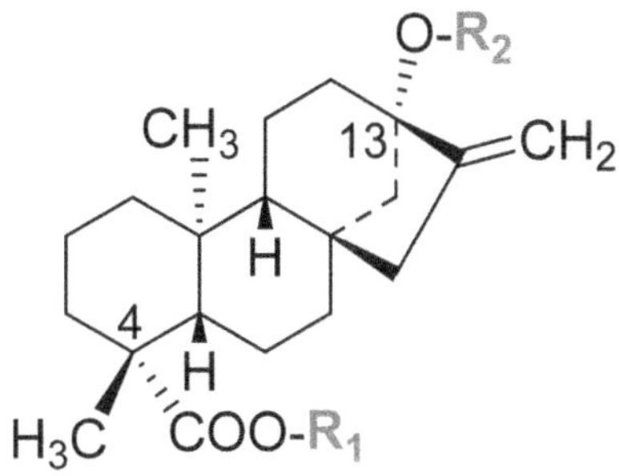

Figure 4.1 Backbone structure of steviol glycosides, R_1 and R_2 are described in Table 4.1.

been isolated from commercially available stevia leaf extract with modification in the *ent*-kaurene skeleton. *Ent*-kaur-15-en-19-oic acid, 16ß-hydroxy-*ent*-kauran-19-oic acid and 13-methyl-16-oxo-17-nor-*ent*-kauran-19-oic acid have been classified as modified backbone.

The first official protocol for the analysis of steviol glycosides was published in 1993 by the Japanese Food Additive Association (JAFA) with the four sweet components stevioside, rebaudioside A and C, and dulcoside A. This mixture was called the "stevia extract".[11] The forced research on stevia in Japan was due to the ban of artificial sweeteners in the country. Therefore, steviol glycosides were introduced in the production of salted food (*e.g.* soy sauce) where combinations of glycyrrhizin and steviol glycosides or enzymatically modified stevia extracts were used.[12] In this official protocol, high-performance liquid chromatography in combination with ultraviolet detection (HPLC-UV) was applied. Separation occurred on an amino column with isocratic elution and was the basis for the first method published by the Joint FAO/WHO Expert Committee on Food Additives (JECFA) in 2008 considering eight steviol glycosides.[13]

In 2010, JECFA revised the method and introduced a reversed phase column instead of the amino column and expanded the number of steviol glycosides to ten components (see Figure 4.1 and Table 4.1).[14] All these official methods were developed for the analysis of high purified steviol glycoside preparations, where more than 80% of them presented 95% total steviol content. Steviol glycosides were introduced as food additives on the European market again with a content of more than 95% steviol glycosides in 2011 considering 11 steviol glycosides with rebaudioside E included (Figure 4.1 and Table 4.1).[15] This was finally extended to 12 steviol glycosides in 2016 with the addition of rebaudioside M.[16] To date, no official protocol has been published for the analysis of these components. So far, this is a critical point as EU regulations have set up maximum levels for 31 different food categories that cannot be controlled by a validated protocol.[17]

A lot of publications pay attention to the analysis of stevia leaves as the total amount and the composition of steviol glycosides vary considerably due to several influences like genotypes, climate and growth conditions. These protocols are often used to control the success of cultivation

Table 4.1 Steviol glycosides (see Figure 4.1) being represented in different official methods or regulations, JAFA = Japanese Food Additive Association, JECFA = Joint FAO/WHO Expert Committee on Food Additives, EU = European Commission Regulation (Glc = glucose, Rha = rhamnose, Xyl = xylose).

Name	R_1	R_2	JAFA 1994[11]	JECFA 2008[13]	JECFA 2010[14]	EU 2012[15]	EU 2016[16]
Steviol	H	H		●	●	●	●
Rubusoside	Glcβ1-	Glcβ1-		●	●	●	●
Steviolbioside	H	Glcβ1-2 Glcβ1-		●	●	●	●
Dulcoside A	Glcβ1-	Rhaα1-2 Glcβ1-	●	●	●	●	●
Stevioside	Glcβ1-	Glcβ1-2 Glcβ1-	●	●	●	●	●
Rebaudioside A	Glcβ1-	Glcβ1-2 (Glcβ1-3) Glcβ1-	●	●	●	●	●
Rebaudioside B	H	Glcβ1-2 (Glcβ1-3) Glcβ1-		●	●	●	●
Rebaudioside C	Glcβ1-	Rhaα1-2 (Glcβ1-3) Glcβ1-	●	●	●	●	●
Rebaudioside D	Glcβ1-2 Glcβ1	Glcβ1-2 (Glcβ1-3) Glcβ1-			●	●	●
Rebaudioside E	Glcβ1-2 Glcβ1-	Glcβ1-2 Glcβ1-				●	●
Rebaudioside F	Glcβ1-	Xylβ1-2 (Glcβ1-3) Glcβ1-			●	●	●
Rebaudioside M	Glcβ1-2 (Glcβ1-3) Glcβ1-	Glcβ1-2 (Glcβ1-3) Glcβ1-					●

experiments, *e.g.* the influence of nitrogen fertilisation, stress factors or to clarify the different steps in steviol glycoside biosynthesis.

Other papers deal with the metabolic fate of steviol glycosides, their degradation in animals and the human body and the formation of the aglycon steviol. A few papers take a look at the determination of steviol glycosides in food in order to survey their stability in processed food or the compliance of their content with the regulated maximum levels.

In this chapter, the different analytical techniques for the quantification of steviol glycosides are discussed by means of an extended review of the scientific literature concerning every step of the analysis.

4.2 Sample Preparation

Sample preparation is a very important tool for successful determination and quantification in analytical chemistry. The more selective and sensitive the separation and detection is, the less lavish sample preparation can be, *e.g.* a mass spectrometric (MS) detector can be successfully applied without difficult sample preparation.

4.2.1 Isolation of Steviol Glycosides from Leaves

The sample clean-up of stevia leaves is well established and has been reported for many years. Steviol glycosides from dried leaves are usually obtained after hot water leaching sometimes without further clean-up steps.[18–23] This is not problematic if the most abundant steviol glycosides (stevioside and rebaudioside A) are analysed but if minor components have to be quantified the high matrix load can disturb a satisfactory separation and detection. To remove these disturbing matrices, some authors published protocols with an additional solid phase extraction (SPE), mainly with C18 cartridges.[24–28] Figure 4.2 illustrates the influence of SPE on the detection of steviol glycosides.

Due to the high matrix load visible in the beginning of the chromatogram (see Figure 4.2b) the detection of steviol glycosides with shorter retention times than stevioside, such as steviolbioside, rubusoside and dulcoside A is only possible if SPE has been applied (see Figure 4.2a).[26] These interfering components are other constituents in the leaves such as proteins, fat, polyphenols, fibre *etc.*

Lorenzo *et al.* clarified the water extracts of unmilled dried stevia leaves by ultrafiltration with 3 and 5 KDa membranes.[21] With membranes of 3 KDa they could remove most of the impurities disturbing the following HPLC analysis but the recovery was only 50%. This was in accordance to Chhaya *et al.* obtaining a recovery of approximately 45%.[29]

Many researchers looked for alternative extraction methodologies in order to shorten extraction time, to reduce extraction temperature or to avoid organic solvents. Among the different alternative procedures: ultrasounds, microwaves, pulsed electric technologies, sub and supercritical fluid, and pressurised liquid extraction have been exploited. Jaitak *et al.* compared different

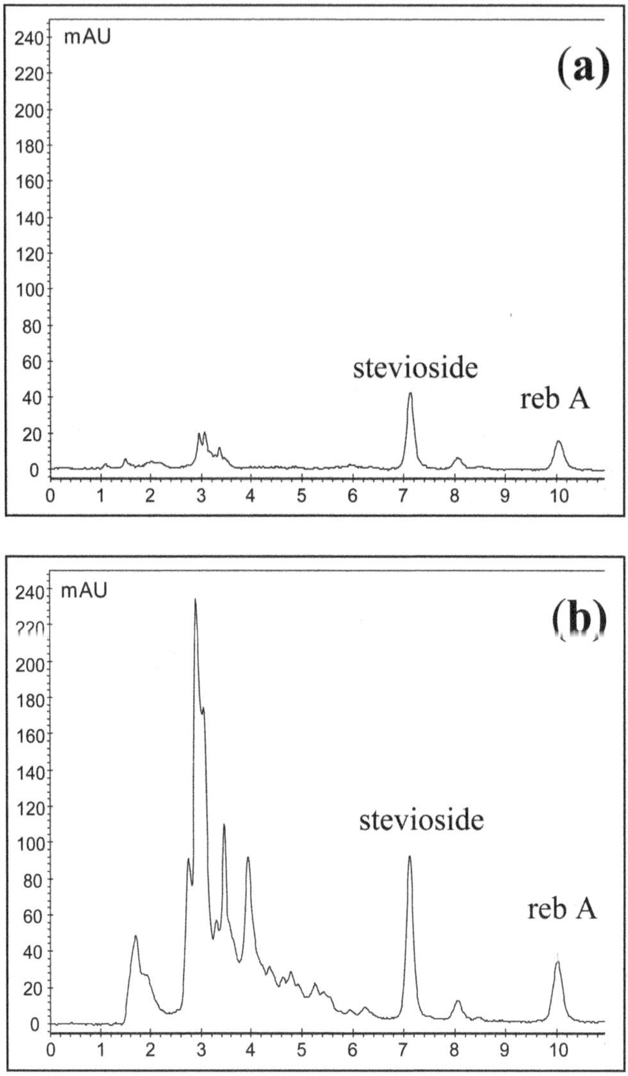

Figure 4.2 Chromatograms of stevia leaf extracts (a) after and (b) without SPE clean-up according to Wölwer-Rieck *et al.*[26] under HILIC conditions and UV-detection at 210 nm.

extraction techniques for stevia leaves.[30] Conventional cold extraction was performed at 25 °C for 12 h, ultrasound-assisted extraction (UAE) at 35 ± 5 °C for 30 min and microwave-assisted extraction (MAE) at 50 °C for 1 min. The authors could show that MAE in combination with the solvent mixture methanol–water (8 : 2 v/v) afforded the highest yields of stevioside and rebaudioside A followed by conventional extraction and the ultrasound method.

Liu *et al.* applied the UAE technique and could confirm an increase of 19% in rebaudioside A and 43% in stevioside yields, respectively.[31] Alapului *et al.* reported a reduction in extraction time compared to conventional maceration from 24 h to 20 min.[32] Further authors reported similar improvements with MAE and UAE technologies.[33–37]

Pressurised fluid extraction has been tested with methanol and pure water. With both solvents stevioside could be extracted with comparable yields.[38]

Traditional methods with chloroform and ethanol as extraction solvent were applied as well as supercritical fluid extraction (SFE). The use of CO_2 and either water or different mixtures of ethanol–water or methanol–water as co-solvents in SFE has been studied by several research groups.[39–41]

In liquid chromatographic mass spectrometry (LC-MS), difficult sample preparation is normally not required.[42] The leaves were extracted with water, water–ethanol or water–methanol[43–47] and after centrifugation the supernatant was diluted and analysed directly.

It is also possible to analyse stevia leaves without any sample preparation using desorption electrospray ionisation mass spectrometry (DESI-MS).[48] The semi-quantitative experiments were performed directly on leaf fragments after an internal standard was spotted on the leaf and allowed to dry before analysis. The components of interest were inserted into the ion source by applying an aqueous spray solution.

4.2.2 Isolation of Steviol Glycosides from Stevia Extract Preparation and Food

Analysis of steviol glycosides in high purity stevia leaf powders, table-top formulations, liquid dietary supplements and beverages can be done after dissolving the samples in water or mixtures of water with methanol, and/or acetonitrile followed by appropriate dilution.[20,43,49–56]

For food samples, a pre-treatment with SPE is often necessary. The authors used C18 cartridges and applied them to soy sauce, candies and pickled radish,[57] orange juice,[24] fruit teas,[25] biscuits, cooked carrots, pickled gherkins,[58] soft drinks,[51] milk products, soy drinks, ice cream, yoghurts, dry biscuits, jam[59] and coffee drinks containing milk.[55] The recoveries for the analysed steviol glycosides were up to 100%.

A modified SPE called dispersive-SPE, was presented by Zhao *et al.*[60] An amino-functionalised magnetic polymer (TEPA-MP) was synthesised and applied to different wine samples spiked with nine food additives including stevioside. It could be shown, that this new solid phase can eliminate most of the interfering matrix components from wine and allows the analysis of stevioside with a LOD of 0.3 and LOQ of 0.99 mg L^{-1}, respectively.

4.3 Separation of Steviol Glycosides by High-performance Liquid Chromatography (HPLC)

HPLC is the method of choice for the determination of steviol glycosides. As they are polar substances due to the attached sugar moieties, they can be analysed without derivatisation. The determination *via* gas chromatography

(GC) is also possible but more laborious and time-consuming as steviol glycosides have to be derivatised to volatile components.

The first chromatographic analyses with HPLC were produced on amino-based and/or reversed phase columns in combination with UV-detection.

4.3.1 Separation on Normal Phase Columns

4.3.1.1 Amino Columns

Amino-based stationary phases show high selectivity for all steviol glycosides and provide good separation of the most abundant isomers rebaudioside B/stevioside and rebaudioside A/rebaudioside E. The separation order predominately depends on molecular weight and therefore on polarity. The more glucose units are attached to the *ent*-kaurene backbone, the higher is their retention time on the column. Accordingly, stevioside (three glucose moieties) elutes before rebaudioside A (four glucose moieties); both are well-separated. Amino-bonded columns were developed by Japanese researchers in the 1970s and 1980s[7,57,61] and found their way into the official protocol for steviol glycoside analysis with four steviol glycosides being separated after isocratic elution.[11] The Joint FAO/WHO Expert Committee on Food Additives (JECFA) published a protocol in 2008 using the same conditions as described in the Japanese official method.[13] Seven steviol glycosides (stevioside, rebaudioside A, rebaudioside B, rebaudioside C, dulcoside A rubusoside and steviolbioside) were analysed and each steviol glycoside could be identified by calculating its retention time relative to rebaudioside A. Both methods were developed for the analysis of highly purified steviol glycosides preparation and not for leaf extracts. Nevertheless, other researchers could show that this column is also suitable for the detection of steviol glycosides in leaf extracts and food, respectively.[18,57,61–65] Makapugay *et al.* analysed eight steviol glycosides in the leaves using gradient elution and Kitada *et al.* could quantify stevioside, rebaudioside A and C and dulcoside A in beverages, soy sauce, candy and pickled radish.[57,62]

Unfortunately, amino-based columns suffer from poor reproducibility and long equilibration times and they cannot be used in combination with MS detectors due to their strong bleeding.

4.3.1.2 Separation on Hydrophilic Interaction Liquid Chromatography (HILIC) Columns

In recent years, many new columns have come onto the market, operating in both the normal and reversed-phase mode. Interesting examples are HILIC columns. They are silica-based, like reversed-phase columns, but possess quite different properties due to functional groups bridging the silica strains by amides, diols, ethylene, ammonium-sulfonic acid or phosphorylcholine ligands.[66–70] Typically, mobile phases comparable to those used on RP columns are applied containing mixtures of organic solvents like acetonitrile and water or aqueous buffers. The retention mechanism in HILIC has received considerable attention in the literature. It is discussed that a water-enriched layer is built up on the surface of the polar stationary phase due to the functional groups which allow partitioning and retention of solutes between the more organic mobile phase and the water layer.[71] Obviously, HILIC columns are suitable for the separation of analytes with poor retention on RPs. They are robust, have a short equilibration time, can be run at higher flow rates due to the lower viscosity of the mobile phase and are suitable for MS detection due to low bleeding.

Wölwer-Rieck *et al.* described for the first time an analytical method for the separation of steviol glycosides under isocratic conditions on a HILIC column using a water–acetonitrile mixture (85 : 15 v/v) as the mobile phase.[26] Steviol glycoside retention on HILIC columns is similar to that on amino phases. Thus, retention times increase with the number of monosaccharide units attached to the *ent*-kaurene skeleton and stevioside elutes before rebaudioside A (Figure 4.3). The protocol has been applied to verify stevioside and rebaudioside A concentrations in cultivation experiments in Germany.

Subsequent publications confirmed the suitability of this column for steviol glycoside analysis.[21,54,72–74] Zimmermann *et al.* successfully tested four different HILIC columns with particle sizes between 1.8 and 3 μm.[72] Isocratic elution was sufficient for the separation of up to seven steviol glycosides within eight minutes. The robustness of the separation depended mainly on the ion strength and the aqueous percentage of the mobile phase.

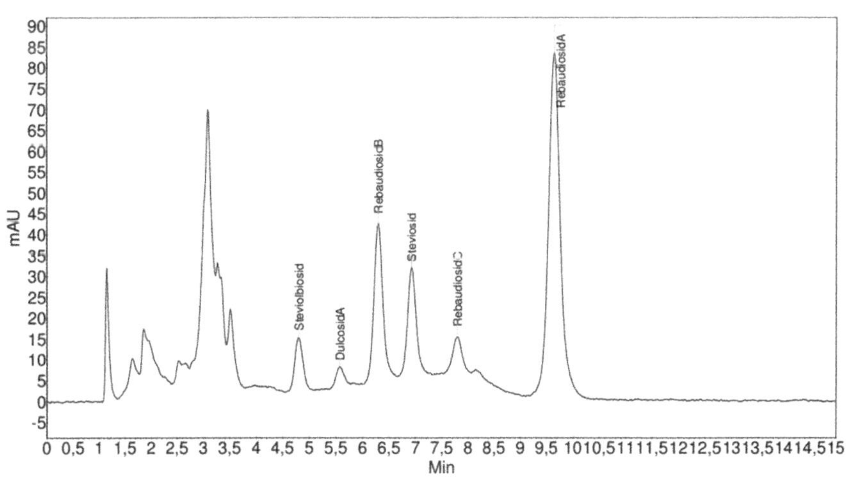

Figure 4.3 Representative chromatogram of steviol glycosides in stevia leaves analysed according to Wölwer-Rieck *et al.*[26] under HILIC conditions and UV-detection at 210 nm.

Nevertheless, HILIC columns are not suitable for the determination of the aglycon steviol. This diterpene is poorly retained on these columns and coelution occurs with some non-specific matrix peaks.[51,74]

4.3.2 Separation on C18 columns

In contrast to amino-bonded and HILIC columns, the retention order is inverted on a C18 phase. Thus, it is possible to detect steviol as well.[27,51] Furthermore, reversed-phase columns are robust and are suitable for MS-detection. In the past, these columns were mainly used for stevioside detection in leaves and food.[24,25,50,75]

In 2009, JECFA required a new protocol as rebaudioside D and F had to be added to the seven already known sweeteners. The JEFCA method of 2008 could not meet the requirement of proper and fast separation of the nine steviol glycosides: rubusoside and steviolbioside were not fully separated and rebaudioside D eluted after 40 min from the amino-based column with a broad pattern and low sensitivity. Moreover, Tada *et al.* confirmed that amino-bonded columns have a tendency to rapidly deteriorate, which makes them unsuitable for a standardised method.[53] Therefore, JECFA revised their HPLC-UV method in 2010 with the introduction of a C18 column and isocratic elution.[14] The mobile phase consisted of a mixture of acetonitrile and 10 mmol L^{-1} sodium phosphate buffer (pH 2.6) (32 : 68 v/v). Under these conditions it was possible to separate the nine steviol glycosides within 25 min. Moreover, each steviol glycoside was identified by an individual reference and each steviol glycoside except rebaudioside A was quantified using a stevioside standard. Rebaudioside A concentrations were calculated on the basis of a separate standard.

One drawback of the JEFCA method is the critical separation of stevioside and rebaudioside A on the C18 columns. Their insufficient resolution ($R < 1.5$) has been criticised and, as only standard solutions have been tested, the limited use of the protocol in the analysis of commercial products or leaves.[76] But recently, methods have been published overcoming these problems due to improved stationary phases or gradient elution.[28,53,76–80] Tada *et al.* addressed these problems by testing several C18 columns and several mobile phases.[53] The best mobile phase was the already mentioned mixture of acetonitrile and water with 10 mmol L^{-1} sodium phosphate buffer (pH 2.6) (32 : 68 v/v). Six different C18 columns were tested but only four showed the required resolution for rebaudioside A and stevioside ($R > 1.5$). Chaturvedula *et al.* could demonstrate the influence of temperature on the resolution and suggested a temperature of 40 or 60 °C for ideal separation.[80]

Bililign *et al.* evaluated the JECFA protocol of 2010 by applying gradient elution. The mobile phase consisted of solvent A (acetonitrile) and solvent B (5 mmol L^{-1} potassium dihydrogen phosphate buffer (pH 3.0)) with a flow rate of 0.5 mL min^{-1}. A specific octadecyl-silica column was used at 32 °C (Figure 4.4).[76] To properly identify the single steviol glycosides,

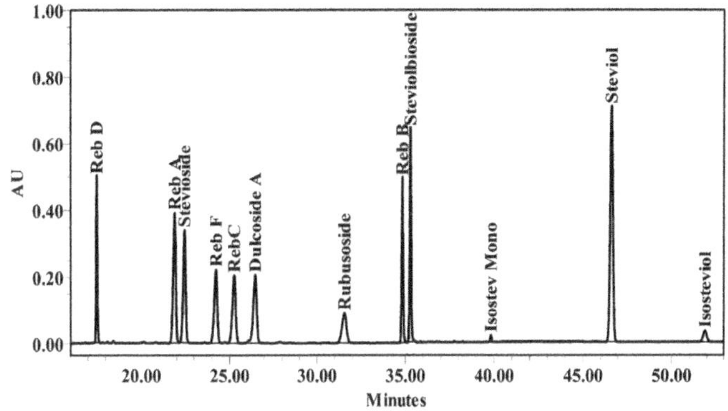

Figure 4.4 Representative chromatogram of several steviol glycoside standards including steviol and iso-steviol analysed according to the protocol of Bililign *et al.*[76] under reversed-phase conditions and UV-detection at 210 nm. Reprinted with permission from ref. 76, Copyright 2014 American Chemical Society.

the United States Pharmacopeia (USP) system suitability reference standard was prepared, which contains the nine steviol glycosides described by JECFA.[14] Quantitative measurements have been done by calculating a relative response factor and validation by applying the USP standard validation procedures.[81] According to Bililign "this validated protocol was found to be suitable to be included by USP as a *Food Chemicals Codex* compendial standard for steviol glycosides".[76]

Martono *et al.* could improve the resolution in the isocratic method of Bergs by adding trifluoroacetic acid (0.01 v/v) to the mobile phase containing a mixture of acetonitrile, methanol and water.[23,28] A good resolution ($R > 2.0$) could be achieved in leaf samples and low calorie candies.

To overcome the problems with the insufficient resolution of stevioside and rebaudioside A some authors proposed to place two C18-columns in series each with a length of 20 or 25 cm, respectively.[59,82] By applying gradient elution with acetonitrile and 1 mmol phosphoric acid nine steviol glycosides could be separated within 20 min.

Nowadays, stationary phases with smaller dimensions and particle sizes have been introduced on the market. They provide a significant reduction in solvent consumption and retention time without losing selectivity. They are often used in ultra-high-performance liquid chromatography (UHPLC), which is generally coupled with MS detectors.

After testing several sub 2-µm and UHPLC columns, Gardana *et al.* achieved the best performance on a 1.8-µm C18 column (150 × 2.1 mm I.D.).[27] They were able to separate and quantify eight known and two unknown steviol glycosides but not steviol within three minutes.

Pól *et al.* used an RP amide C18 column (150 × 2.1 mm I.D.) and a binary gradient with an acetonitrile–water mixture.[83] Wang *et al.* quantitatively determined five steviol glycosides in a 12 minutes run using a 1.8 µm C18 column (100 × 2.1 mm I.D.) and a mobile phase with formic acid (0.05% v/v) for gradient elution.[56] Additionally, they developed a UHPLC-UV method to determine glucose as adulterant in commercial stevia products after 1-phenyl-3-methyl-5-pyrazolone (PMP) derivatisation and could detect glucose in two commercial products with high levels. Zimmermann tested nine state-of-the-art RP columns for the separation of the nine steviol glycosides described in the JECFA method.[14] Two columns could be recommended which allowed the separation within 11 min.[84]

4.3.3 Separation on Further Columns

Some papers describe the use of a hydro reversed-phase column.[85–87] This column combines extreme hydrophobic selectivity with polar end-capping and is suitable for strong non-polar and polar compounds, respectively. For the separation of polar compounds, it is possible to use, if desired, a 100% aqueous mobile phase without causing the collapse of the stationary phase. Hoekstra *et al.* applied a linear gradient containing 1% tetrahydrofuran in water and acetonitrile.[85] He could separate 12 steviol glycosides in a standard solution, including steviol and isosteviol. Clos and Chaturvedula used a gradient of a three-solvent mobile phase system (ammonium acetate–acetic acid buffer, acetonitrile and acetic acid).[86,87] Clos *et al.* separated four steviol glycosides and five degradation products while Chaturvedula *et al.* succeeded in detecting up to 26 steviol glycosides in six leaf extracts of different origin.

Ahmed and Dobberstein used two Protein I-125 columns in series and could detect and quantify eight steviol glycosides.[88,89]

A mixed-mode silica-based column combining hydrophobic and weak anion-exchange properties due to long alkyl chains on its surface was used for the separation of eight steviol glycoside standard solutions.[90] Jaworska *et al.* applied this type of column to plant material and liquid and powdered dietary supplements but the authors recognised that the peaks of rebaudioside A and C and rebaudioside B and D, respectively, were not completely separated under the applied conditions.[20]

4.3.4 Separation Using Two-dimensional Systems

Separation problems can be solved by using two-dimensional systems. In LC × LC systems, the total effluent of the first column is transferred to the second one. This is in contrast to LC–LC systems with multidimensional heart-cutting separation techniques, in which only a selected fraction of the first column is transferred to the second dimension. Cacciola *et al.* used a LC × LC system with a normal phase (polyamine) in the first dimension and a C18 column with sub 2 µm particles in the second dimension and they successfully separated ten steviol glycosides.[19]

Pól combined a small C18 column with an amino-bonded column, both with 3 µm particle size and could quantify nine steviol glycosides in leaf samples.[83]

Cabooter *et al.* developed a LC–LC method on two C18 columns (2.1 × 100 mm I.D.) with particle sizes of 1.7 μm and could successfully separate 26 stevia derived components using an automatic column coupling device. Only eight of these 26 peaks could be assigned to known steviol glycosides at that time.[91]

An offline combination of RP 18/HILIC columns enables the preparative separation of 13 steviol glycosides in leaves.[92]

4.4 Detection in HPLC

4.4.1 UV-detection

UV- and diode array-detection are mostly used in steviol glycoside analysis as it is easy to apply, inexpensive and robust. The detection wavelengths are set at 210 or 200 nm, which normally allows a selective and sensitive detection over a wide concentration range. However, as steviol glycosides are weak UV absorbers (due to the lack of chromophores like conjugated double bonds), their detection is not very sensitive. This is the main disadvantage of UV detection in steviol glycoside analysis. They can only be analysed selectively if the single steviol glycosides are well separated from each other and from disturbing matrix components. Therefore, a lot of effort has to be put into selecting the right column and developing an effective sample preparation.

However, there is an important time-saving advantage in applying UV detectors. All sweet components are detected with the same sensitivity will say have the same response in UV. This is because only the *ent*-kaurene skeleton is responsible for UV absorption, the corresponding sugar moieties show nearly no UV-activity. Thus, a number of steviol glycosides can be quantified by using only one or two standard substances, preferably stevioside and/or rebaudioside A, in high purity and their resulting calibration curves. Further components can be analysed applying factors considering the formula weight. This procedure has been introduced in the validated protocols of JECFA.[13,14]

With UV-detectors the limits of detection (LOD) respective the limits of quantification (LOQ) range in the μg mL^{-1} region. Table 4.2 gives an overview about HPLC methods and published LODs and LOQs.

4.4.2 MS-detection

Mass spectrometry is the most sensitive detection mode for steviol glycosides. There are a lot of different mass spectrometric detectors on the market. Mainly, triple quadrupole mass spectrometers (MS/MS) are used as this technique is very selective. It provides information about the molecular weight as well as about the typical fragmentation of the analyte.

The first quadrupole filter (Q1) selects the mass of the intact analytes (precursor ions), which fragments in Q2 into additional ions (fragments or

Table 4.2 LODs and/or LOQs of steviol glycosides detected by UV or DAD (Stev = stevioside, RebA = rebaudioside A, Rub = rubusoside, Stevbio = steviolbioside, DulcA = dulcoside A).

LOD (μg mL^{-1})	LOQ (μg mL^{-1})	References
1.5 (Stev)		24,25
50 (Stev)		25
1.25	2.5	82
0.4 (Reb A)	3.8 (Reb A)	28
2 (steviol), 3 (DulcA), 6 (Stev), 4 (Stevbio), 6 (RebC), 15 (RebA), 10 (RebD), 2 (RebB)	10 (DulcA), 20 (Stev), 12 (Stevbio), 20 (RebC), 50 (RebA), 33 (RebD), 5 (RebB)	20
	4 (DulcA), 4 (Rub), 5 (Stevbio), 3 (Stev), 4 (RebC), 3 (RebA), 2 (RebD), 6 (RebB), 4 (RebF)	53
1.1 (Stev, RebA)	3.6 (Stev, RebA)	21
	10	76
8.53 (RebD)	25.85 (RebD)	22
5 (Stevbio), 10 (DulcA), 7 (Stev, RebA, RebD)	15 (Stevbio), 30 (DulcA), 25 (Stev, RebA, RebD)	56
1 (RebA), 1 (Stev)	3 (RebA), 3 (Stev)	23

product ions) by collision with an inert gas, for example argon or nitrogen. These fragments are selected afterwards in the mass filter Q3.

The precursor ions can give hints about the molecular weight of the analyte when a soft ionisation technique is used. The fragment ions often show a typical pattern of fragmentation and can give additional information about the identity and structure of the analyte or even about a class of analytes. When the specific fragmentation pattern of the analyte is known a scan-mode, called multiple reaction monitoring (MRM), can be applied in the detection of two or more specific fragments from a defined precursor ion. This detection method is very sensitive and specific and can be used for quantification.

It is quite challenging to couple the MS-detector to HPLC- or UHPLC-devices. These instruments deliver eluents consisting of aqueous and/or organic solvents, but the MS-detector operates under high vacuum to maintain the typical quadrupole property and requires evaporated and selected charged molecules to be detected and fragmented. All these operations have to be done by a powerful interface between HPLC and MS. It has to evaporate the effluent of the HPLC column, to volatilise and ionise the remaining analyte molecules and to transfer a part of them as ions into the MS. The most commonly used interfaces are electrospray ionisation (ESI) and atmospheric pressure chemical ionisation (APCI).[71] They differ in the kind of ionisation. In APCI, the analytes are charged after evaporation in the gas phase by applying a corona-discharge needle. APCI is used for compounds which do not ionise very well due to high stability, low polarity or small molecular weight. In ESI, the analytes are charged before evaporation by a potential which is placed on the entrance nebuliser spray tip. ESI ionisation is said to be a softer ionisation technique and therefore often used for bioanalytical applications. In both interfaces a positive or negative potential can be applied depending on the property of the analytes.

For steviol glycoside analysis, the ESI interface is frequently applied in combination with positive and negative ionisation. Authors who compared the two modes suggested the negative mode.[39,46,93–95] Choi *et al.* stated that negative ion mode is ten times more sensitive than the positive ion mode.[39] Molina-Calle *et al.* also registered a better sensitivity. Koyama *et al.* showed that exclusively in the negative mode steviol and two metabolites could be detected in human faeces.[46,95]

Mobile phases contain acetonitrile–water or methanol–water mixtures and additives like volatile acids, bases or buffers. The additives promote ion formation will say improve the ionisation of the analytes, which enhances the sensitivity of the detection. Ammonium acetate,[27,43,44,73,74,93–95] ammonium formate or formic acid,[46,47,55,72,96,97] trifluoroacetic acid[45] or dichloromethane[27,98] are frequently used as volatile additives.

In the positive ion mode the precursor ions are often presented as adducts of sodium $[M + Na]^+$, potassium $[M + K]^+$ or ammonium $[M + NH_4]^+$ (for these ions the term "pseudo-molecular ions" is still used although called deprecated by IUPAC).[99]

In negative ion mode the precursor ion represents the molecular weight of the molecule minus a proton $[M - H]^-$.

The fragmentation pattern of steviol glycosides in LC-ESI-MS is rather identical for all sweet molecules. The subsequent loss of fragments with 162 Da represents the loss of one glucose moiety and the loss of 146 Da the loss of one pentose (rhamnose or xylose) moiety, which is typical for rebaudioside C and F and dulcoside A.

Table 4.3 gives an overview of the representative precursor and fragmentation ions of the 11 most abundant steviol glycosides. These typical

Table 4.3 Precursor and typical fragment ions in HPLC-ESI-MS/MS, negative ionisation mode, the most abundant ion in bold are used for MRM-transition measurements.

Component	Molecular weight	$[M - H]^-$ (*m*/*z*)	Main fragment ions (*m*/*z*)
Steviol	318	317	
Rubusoside	642	**641**	**479**, 317
Steviolbioside	642	**641**	**479**, 317
Dulcoside A	788	**787**	**625, 479**
Reb B	804	**803**	**641**, 479
Stevioside	804	803	**641, 479**
Reb C	950	**949**	**787**, 625
Reb A	966	965	**803, 641**
Reb F	936	**935**	**773, 641**
Reb D	1128	**1127**	**803**, 641
Reb E	966	965	**803, 641**

fragment ions can also be used for quantification in MRM mode. Table 4.4 presents LODs and LOQs in the ng mL^{-1} range. Obviously, MS-detectors are about 1000 times more sensitive than UV-detectors for steviol glycoside quantitation.

Due to the uniform fragmentation pattern it is rather difficult to distinguish between isomers such as rubusoside and steviolbioside, stevioside and rebaudioside B or rebaudioside A and E especially when LC resolution is not sufficient. Some authors confirmed that a distinction is possible when applying low, intermediate, and high collision energies (10–100 eV) in MS-detection.[97,100–102] They could prove that the ester bond between the glucose moiety and the carboxyl group at C4 (C19) of the *ent*-kaurene backbone (acylglycoside isomers) fragmented quite easily, even if low collision energies were applied, while higher collision energy has to be applied to cleave the sugar bond at C13. Thus, the precursor ions had low intensities or were even absent in the MS^2 spectra. Although ESI is a soft ionisation technique, the sugar moieties at C19 (carboxyl group) can get lost in the interface. This phenomenon is called in-source fragmentation and leads to ions with lower *m/z* values in the MS spectra than expected for the $[M - H]^-$ ion. For example, the $[M - H]^-$ ion of stevioside (*m/z* 803) was very low but the $[M{-}Glu]^-$ peak (*m/z* 641) appeared with high intensity.[44,46,73,97,102] Due to this easy fragmentation, the corresponding steviol glycosides (*e.g.* stevioside, rebaudioside A, dulcoside A) can unambiguously be confirmed by the relative intensity of their fragment ions. The bond of glucose at C13 is more stable and the resulting steviol glycoside (*e.g.* steviolbioside, rebaudioside B) have stable $[M - H]^-$ ions even at higher voltage settings. This could also be proven in an ESI-QTOF-MS/MS experiment with direct infusion of the samples.[102] The authors could show that even a low collision energy of 10 eV can cause the cleavage of a monosaccharide at C19. They even established universally applicable dissociation patterns for steviol glycosides with di- or trisaccharides attached at C13 and/or C19. This is of high importance for rapid detection and identification of known and new steviol glycosides.

Time-of-flight mass spectrometers (TOF-MSs) operate in a different way than quadrupole detectors. Here, the velocity of ions generated in an interface is the important parameter. The ions are accelerated by a specific energy and directed through a drift tube. The higher the mass of the ions the slower they are drifting through the tube. Therefore, the time for travelling through the tube correlates to the mass of the ions. TOF-MSs provide higher mass resolution, are more sensitive but not preferable for quantitation.[48,52,83,100] Pol *et al.* applied this detector to leaf samples, and Kakigi *et al.* could detect up to 30 steviol glycosides and applied the protocol to nine different soft drinks.[52,83]

Quantification with MS-detectors in complex samples like food is challenging because matrix effects can influence the performance of the analytical method. This phenomenon is called ion suppression or

Table 4.4 Ionisation modes and LOD/LOQs in LC-ESI-MS/MS for several steviol glycosides (Stev = stevioside, Reb A–F = rebaudiosides A–F, Stevbio = steviolbioside, Rub = rubusoside, DulcA = dulcoside A, Stevmono = steviolmonoside) in different matrices.

Components	Matrix	Ionisation mode	LOD	LOQ	References
Stev	Stevia leaves	Negative + positive			39
Stev, Reb A, C, DulcA, Rub, 5 metabolites, steviol	Human intestinal microflora	Negative + positive			95
Stev, Reb A, C, Rub, Stevbio, Stevmono	Stevia leaves	Negative			100
Stev, Reb A, C, DulcA	Soju		75 μg mL^{-1} (Stev)		42
Stev, Reb A–E, Stevbio, DulcA	Plant material, suspension culture	Positive			127
Stev	Several sweeteners in food	Negative	0.1 μg mL^{-1}	0.3 μg mL^{-1}	128
20 steviol glycosides	Stevia leaves	Negative			101
Stev, Reb A–C, E, Rub, Stevbio, DulcA, steviol, 2 unknowns	Stevia leaves, purified stevia extracts	Negative	1–50 ng mL^{-1}		27
Steviol, Stev, Reb A	Soft drinks	Negative	6 ng mL^{-1} (steviol)	60 ng mL^{-1} (steviol)	51
Stev, Reb A–C, Rub, Stevbio, DulcA	Highly purified stevia extract	Negative			72
Stev, Reb A–F, Rub, Stevbio, DulcA, 12 unknown	Highly purified stevia extract	Negative			97
Stev, Reb A-F, Rub, Stevbio, DulcA, steviol	Sweetener products	Negative	0.01–0.34 μg g^{-1} powder		43
Stev, steviol	Juices, coffee	Positive			129
Steviol, iso-Stev, Stevbio	Standards	Positive + negative			130
Stev, Reb A–C, Rub, Stevbio, DulcA, steviol, iso-Stev	Stevia leaves	Negative	0.7–14 ng mL^{-1}	2.3–47.6 ng mL^{-1}	44
Stev, Reb A–D, F, Stevbio	Plant material	Positive	0.1–2.8 ng mL^{-1}	2.1–8.9 ng mL^{-1}	45
Stev, Reb A–F, Rub, Stevbio, DulcA, dihydro-components	Stevia leaves	Negative			73

(*continued*)

Table 4.4 *(continued)*

Components	Matrix	Ionisation mode	LOD	LOQ	References
Stev, Reb A, C, Rub, Stevbio, DulcA	Purified extracts	Negative			98
Reb A	Beverages with other sweeteners	Negative	5 ng mL^{-1}	20 ng mL^{-1}	55
Stev, Reb A–D, Stevbio, DulcA, steviol	Beverages with other sweeteners	Negative			74
Stev, Reb A–D, Rub, Stevbio, DulcA	Stevia leaves	Negative + positive	0.1–0.5 ng mL^{-1}	0.5–1 µg mL^{-1}	46
Stev, Reb A–D, Stevbio, DulcA, steviol	Stevia leaves	Positive	1.1–1.8 ng mL^{-1}	2.1–6.7 ng mL^{-1}	47

promotion of ionisation[55] and may cause false results. To evaluate these effects, external calibration curves have to be compared to standard addition curves for every type of matrix. Sakai *et al.* could show that matrix effects were negligible in different types of beverages and cocktails.[55] They recommended to apply the standard addition method to samples with high matrix content, such as *e.g.* coffee with milk, peanut butter, jam *etc.* They confirmed the results of Shah *et al.* who could quantify the steviol glycoside content in a chocolate drink correctly after standard addition calibration.[43]

An alternative approach to improve precision in complex matrices is the application of stable isotope dilution analysis (SIDA). The principle is as follows: an isotope labelled internal standard (IS) *e.g.* a deuterium-labelled standard has been added to the sample at the beginning of the whole analytical procedure and after its equilibration with the analyte passes through all subsequent steps until detection. The ratio of the isotopologues remains stable during the whole analytical procedure. This is due to their almost identical chemical and physical properties. Differentiation occurs finally in the MS detector as both the analyte and the labelled standard differ in their molecular weight and can be easily distinguished by their different masses.[103,104]

The main disadvantage of this calibration technique is the fact that labelled standards have often to be synthesised as they are usually not commercially available.

Well *et al.* applied this technique to the quantitation of 10 steviol glycosides in foods and beverages using the corresponding deuterated 16,17-dihydrosteviol glycosides as internal standards.[73]

4.4.3 Applying Alternative Detectors

An alternative for UV-detection is the use of an evaporative light scattering detector (ELSD) or a charged aerosol detector (CAD). Both can be called universal or mass-dependent detectors.

Their detection principles are similar: at first, the mobile phase of the chromatographic system is nebulised by a flow of nitrogen. The resulting aerosol is transported through a tube where the volatile components and solvents are evaporated; only the non-volatile molecules remain and are detected. In ELSD, the molecules are measured by their light scattering with a photodetector, in CAD they are mixed with a stream of nitrogen that has been positively charged by a corona-discharge device, which is a high-voltage platinum wire. This charge is transferred to the analyte particles which generate a signal that is read by an electrometer.[71,105]

It is evident for both detectors, that the mobile phase is restricted to volatile components (no buffers with inorganic salts). The sensitivity of the ELSD in steviol glycoside analysis is comparable to that of a UV-detector but allows the detection of additional unknown peaks.[90,106] The linear range of calibration curves is more restricted in ELSD. Wang *et al.* overcame this problem by applying logarithmic calibration, which shows a linear correlation between responding peak areas and amount of analytes.[56]

With CAD, the sensitivity can be raised by a factor of 3–5.[86,87]

Clos *et al.* used this detector to identify degradation products of rebaudioside A lacking olefinic unsaturation in the steviol backbone.[86]

Furthermore, sensitive detection of stevioside is possible using a pulsed amperometric detector (PAD).[64] After HPLC separation, the effluent was adjusted to pH ≥ 12 using sodium hydroxide and after passing through a reaction coil, stevioside was quantified *via* its glycose moieties in the amperometric detector with a LOD of 0.3 mg L^{-1} for standards. The protocol was successfully applied to the analysis of stevia leaves.

4.5 Alternative Techniques for Steviol Glycosides Determination

4.5.1 Thin Layer (TLC) and High Performance Thin Layer Chromatography (HPTLC)

Thin layer chromatography (TLC) has been used since the beginning of steviol glycoside analysis.[65,107–110] It still has a place today as the technique is "efficient in terms of time, yield and it is also simple, specific, rapid, reproducible and cost effective".[111] By the introduction of high-performance thin-layers (HPTLC) the technique has become an alternative to HPLC if a high sample throughput for routine screening is required. Pre-coated silica gel 60 F_{254} plates are normally used, mobile phases are of different compositions and densitometric detection and quantification is done after visualisation of the steviol glycosides with several detection reagents.

Table 4.5 gives an overview of the detectable steviol glycosides, the analysed matrices and the achieved LODs and LOQs of this technique. The most comprehensive paper in HPTLC analysis was published by Morlock *et al.*[54] The protocol was applied to sugar-free food products and stevia formulations. The quantification of rebaudioside A in stevia formulations was compared to HILIC-HPLC and RP-UHPLC methods and it could be shown that HPTLC is a powerful alternative with regard to high sample throughput capacity as it is less expensive and much faster but provides comparable results after simple sample preparation. The method is limited to the determination of major steviol glycosides in sugar-free foods as high contents of mono and disaccharides disturb the exact determination. Furthermore, a HPTLC-MS coupling was described using a commercially available TLC-MS interface.

4.5.2 Capillary Electrophoresis (CE)

Only a few publications are available using this technique. Liu *et al.* used a borate buffer (pH = 9.2) modified with 45% (v/v) acetonitrile.[112] Mauri *et al.* employed micellar electrokinetic chromatography in the presence of

Table 4.5 HPTLC conditions and LODs/LOQs for steviol glycosides (Stev = stevioside, Reb A–D = rebaudiosides A–D, Stevbio = steviolbioside, DulcA = dulcoside A) in different matrices.

Components	Samples	Mobile phase	Spray reagent	LOD	LOQ	References
Stev, Stevbio, RebA	Stevia leaves	Ethyl acetate–ethanol–water (80 : 20 : 12)	Acetic anhydride–sulfuric acid–ethanol (01 : 01 : 10)	80–120 ng	160–1000 ng	131
Stev, Reb A	Stevia plants	Acetone–ethylacetate–water (5 : 4 : 1)	Anisaldehyde–sulfuric acid	35 ng	100 ng	132
Stev, RebA	Stevia plants	Ethylacetate–ethanol–acetone–water (15 : 3 : 6 : 6)	Anisaldehyde–sulfuric acid	127 (Stev), 393 (RebA) ng	387 (Stev), 1191 (RebA) ng	133
Stev, RebA	Stevia plants	Chloroform–methanol–water (60 : 32 : 4)	Acetic anhydride–sulfuric acid–ethanol (1 : 1 : 4)	100 (Stev), 40 (RebA) ng	200 (Stev), 100 (RebA) ng	111
Stev, Reb A–D, Stevbio, DulcA, Stevbio	Stevia formulation, sugar-free foods	Ethylacetate–methanol–acetic acid (3 : 1 : 1)	β-naphthol	10 ng	30 ng	54

sodium dodecylphosphate and Dacombe *et al.* added γ-cyclodextrin to the buffered eluent, which even allowed the separation of steviolbioside and steviol.[65,113]

Using electrokinetic-CE in the presence of a derivative β-cyclodextrin as separating agent, stevioside and rebaudioside A were detected with LODs of 59.4 and 19.4 $\mu g\ mL^{-1}$, respectively. The method was successfully applied to commercial samples like powders, liquids and tablets.[114] Furthermore, molecular docking studies were carried out to better understand the complexes formed between the two steviol glycosides and the separating agent. Pavlicek *et al.* proposed two different methods with contactless conductivity detection and could analyse rebaudioside A and stevioside with a LOD of 0.3 $mg\ L^{-1}$.[115]

4.5.3 ^{1}H-NMR Spectroscopy

The major components stevioside and rebaudioside A–C could be quantified by ^{1}H-NMR spectroscopy.[116] The solvent mixture pyridine-d_5–DMSO-d_6 (6 : 1 v/v) enables satisfactory separation of the interesting components. Compared with the JECFA method similar results were obtained.[14] NMR analysis offers the advantage that reference substances are not required and that it is significantly faster than HPLC analysis. Tada confirmed the actual contents of stevioside and rebaudioside A in standards by qNMR.[117]

4.5.4 Near-infrared Reflectance Spectroscopy (NIRS)

Another method for rapid quantitation of major steviol glycosides in leaves is NIRS. Reference analysis has been done by HPLC. Hearn *et al.* could analyse stevioside rebaudioside A and C, and Yu *et al.* and Tang *et al.* stevioside and rebaudioside A.[118–120] The protocol was applied to stevia leaf samples and it could be shown that the high correlation between the values predicted by the model and the values analysed by HPLC makes NIRS a powerful tool for rapid analysis of the main steviol glycosides.

4.5.5 Raman Spectroscopy

Jentsch *et al.* showed, by applying Raman spectroscopy, that three Bolivian stevia products were counterfeited with sodium cyclamate and small amounts of sodium saccharin. Sodium saccharin could be detected at a concentration as low as 5% in stevia preparations.[121]

4.5.6 Detection of Steviol Glycosides as Sum Parameter

As the maximum levels of steviol glycosides in food categories are expressed as steviol equivalents, Bartholomees *et al.* developed a method to quantify the sweeteners *via* isosteviol as a sum parameter.[17,122] Their RP-HPLC method with fluorescence detection used the fact that all steviol glycosides are hydrolysed to isosteviol by incubation with diluted sulfuric acid. Isosteviol was converted to a fluorescent derivative and quantified with low LOQs and good reproducibility using the internal standard dihydroisosteviol. The advantage of this protocol is that only one calibration curve has to be established to quantify all steviol glycosides. The authors applied this protocol successfully to commercial steviol glycosides mixtures and different food matrices.

Gas chromatographic analysis in combination with MS detection (GC/MS) of the hydrolysis product isosteviol has been reported in biosynthesis research on steviol glycosides. All steviol glycosides were hydrolysed to isosteviol as described by Nowotny.[123] Isosteviol was converted by *N*-Methyl-*N*-(trimethylsilyl)trifluoroacetamide (MSTFA) to its trimethylsilyl (TMS)-derivative. This volatile compound was injected in the GC and identified in the mass spectrometric detector *via* its typical TMS-ester fragments.[103]

4.5.7 Detection of the Aglycon Steviol

Steviol, the aglycon of steviol glycosides, has often been analysed in toxicological, pharmacokinetic and metabolic studies. In humans, all steviol glycosides are degraded by microorganism to steviol in the colon. Thus, steviol is the only metabolite in human faeces, in the kidney it is metabolised to steviol glucuronide and excreted as such in urine.[124]

Steviol has not been detected in leaves[45] or only in very small amounts (2.7–13.2 mg kg^{-1} and 5.9 ± 0.8 mg kg^{-1}).[27,125] In highly purified powders, food beverages or dietary supplements steviol could not be detected either.[20,51,125,126]

Because of differences in polarity and chemical structure the simultaneous separation of steviol and steviol glycosides is challenging. Minne *et al.* described a simple method by RP-HPLC using dihydroisosteviol (DHISV) as the internal standard. Steviol and DHISV were derivatised with 4-(bromomethyl)-7-methoxycoumarin and the resulting ester derivatives were determined with a fluorescent detector. A LOD of 100 pg per injection could be achieved.[125]

Further protocols analysed steviol without derivatisation using RP columns and UV or more sensitive MS detectors. As steviol is less polar than the glycosylated diterpenes it elutes with very high retention times from RP columns. Therefore, Gardana *et al.* and Wölwer-Rieck *et al.* determined steviol separately by using a mobile phase of different composition. Steviol could be detected in ESI-MS/MS with negative ionisation by the precursor/product ion *m*/*z* 317. No fragment ions were apparent.[27,51]

By applying ESI-MS/MS in the positive mode, fragmentation of steviol could be observed and steviol was detected *via* the precursor ion (*m*/*z* 319.2) and the fragment ion (*m*/*z* 275.0).[45] The authors applied the method to leaves of *Stevia rebaudiana* from different origins.

Jaworska *et al.* succeeded in detecting steviol together with steviol glycosides in one run using a mixed-mode column that exhibits both hydrophobic

Table 4.6 LOD/LOQs in LC-ESI-MS/MS for steviol in different matrices.

Matrix	Ionisation mode	LOD	LOQ	References
Stevia leaves	Negative	2.5 ng g^{-1}		27
Rat liver	Negative	1 ng mL^{-1}		134
Soft drinks	Negative	6 ng mL^{-1}		51
Human faecal homogenate	Negative	50 ng mL^{-1}		135,136
Plasma	Negative		20 ng mL^{-1}	137
Stevia leaves	Positive	1.1 mL^{-1}		45

and hydrophilic properties. Steviol elutes rather early and was detected by UV with a LOD of 2 and a LOQ of 7 μg mL^{-1}, respectively.[20] Wald *et al.* developed a HPTLC method combined with MS-detection in the positive mode that allows the proof of absence for steviol and isosteviol in different stevia formulations and leaves.[126]

Table 4.6 provides an overview of achievable LODs and/or LOQs for steviol in different matrices.

4.6 Conclusions

Steviol glycosides are rather polar and non-volatile due to the sugar moieties attached to the *ent*-kaurene skeleton. Therefore, the analytical methods of choice are based on liquid chromatography. A large selection of separation phases is available, but mostly reversed-phase columns are applied. Detection has often been done by UV-detectors set at a wavelength of 190–210 nm, but this mode is not very sensitive. As mass spectrometric detectors were continually applied in the quantitative analysis of very different components, they were also successfully introduced in steviol glycosides analysis mainly as quadrupole MS with an ESI interface operating in the negative ionisation mode. LOQs and LODs are about 1000 times better compared to those of UV detection. Further techniques like HPTLC, NMR, NIRS and CE can also be used but cannot compete with liquid chromatographic analysis concerning variability, selectivity and sensitivity. Their advantage is their speediness, which makes them suitable for screening analyses.

References

1. J. E. Brandle, A. N. Starratt and M. Gijzen, *Can. J. Plant Sci.*, 1998, **78**, 527.
2. M. Bridel and R. Lavieille, *C. R. Acad. Sci.*, 1931, **192**, 1123.
3. E. Mosettig and U. Beglinger, *et al.*, *J. Am. Chem. Soc.*, 1963, **85**(5), 2305.
4. H. Kohda, R. Kasai, K. Yamasaki, K. Murakami and O. Tanaka, *Phytochemistry*, 1976, **15**, 981.

5. I. Sakamoto, K. Yamasaki and O. Tanaka, *Chem. Pharm. Bull.*, 1977, **25**, 3434.
6. I. Sakamoto, K. Yamasaki and O. Tanaka, *Chem. Pharm. Bull.*, 1977, **25**, 844.
7. M. Kobayashi, S. Horikawa, I. H. Degrandi, J. Ueno and H. Mitsuhashi, *Phytochemistry*, 1977, **16**(9), 1405.
8. S. Purkayastha, A. Markosyan, I. Prakash, S. Bhusari, G. Pugh Jr., B. Lynch and A. Roberts, *Regul. Toxicol. Pharmacol.*, 2016, **77**, 125.
9. M. A. Ibrahim, D. L. Rodenburg, K. Alves, F. R. Fronczek, J. D. McChesney, C. Wu, B. J. Nettles, S. K. Venkataraman and F. Jaksch, *J. Nat. Prod.*, 2014, **77**(5), 1231.
10. M. A. Ibrahim, D. L. Rodenburg, K. Alves, W. H. Perera, F. R. Fronczek, J. Bowling and J. D. McChesney, *J. Nat. Prod.*, 2016, **79**(5), 1468.
11. Japan Food Additive Association, in *Voluntary Specifications of Non-chemically Synthesized Food Additives*, ed. Japan Food Additive Association, Tokyo, 1993, p. 119.
12. K. Mizutani and O. Tanaka, in *Stevia, The Genus Stevia*, ed. A. D. Kinghorn, Taylor & Francis, London and New York, 2002, p. 178.
13. Joint FAO/WHO Expert Committee on Food Additives (JECFA), 69th meeting, *Compendium of food additive specifications: Steviol Glycosides*, FAO JECFA Monographs 5, Rome, 2008.
14. Joint FAO/WHO Expert Committee on Food Additives (JECFA), 73rd Meeting, *Compendium of Food Additive Specifications: Steviol Glycosides*, FAO JECFA Monographs 10, Rome, 2010.
15. Commission regulation (EU) No 231/2012, *Off. J. Eur. Union*, 2012, **L 83**, 1.
16. Commission regulation (EU) 2016/1814, *Off. J. Eur. Union*, 2016, **L 278**, 37.
17. Commission regulation (EU) No 1131/2011, *Off. J. Eur. Union*, 2011, **L 295**, 205.
18. N. Kolb, J. L. Herrera, D. J. Ferreyra and R. F. Uliana, *J. Agric. Food Chem.*, 2001, **49**(10), 4538.
19. F. Cacciola, P. Delmonte, K. Jaworska, P. Dugo, L. Mondello and J. I. Rader, *J. Chromatogr. A*, 2011, **1218**(15), 2012.
20. K. Jaworska, A. J. Krynitsky and J. I. Rader, *J. AOAC Int.*, 2012, **95**(6), 1588.
21. C. Lorenzo, J. Serrano-Diaz, M. Plaza, C. Quintanilla and G. L. Alonso, *Food Chem.*, 2014, **157**, 518.
22. I. Aranda-González, Y. Moguel-Ordoñez and D. Betancur-Ancona, *Am. J. Anal. Chem.*, 2014, **05**(13), 813.
23. Y. Martono, S. Riyanto, A. Rohman and S. Martono, in *Proceedings of the 1st International Conference on Science and Technology 2015 (ICST-2015)*, ed. AIP Publishing, 2015, vol. 1755, 080001.
24. L. Bovanová, E. Brandšteterová and S. Baxa, *ZLUF*, 1998, **207**, 352.
25. T. Vanek, A. Nepovim and P. Valicek, *J. Food Compos. Anal.*, 2001, **14**, 383.
26. U. Wölwer-Rieck, C. Lankes, A. Wawrzun and M. Wüst, *Eur. Food Res. Technol.*, 2010, **231**(4), 581.

27. C. Gardana, M. Scaglianti and P. Simonetti, *J. Chromatogr. A*, 2010, **1217**(9), 1463.
28. D. Bergs, B. Burghoff, M. Joehnck, G. Martin and G. Schembecker, *J. Verbraucherschutz Lebensmittelsicherh.*, 2012, **7**(2), 147.
29. S. C. Chhaya, S. Mondal, G. C. Majumdar and S. De, *Food Bioprod. Process.*, 2012, **90**(C3), 525.
30. V. Jaitak, B. S. Bandna and V. K. Kaul, *Phytochem. Anal.*, 2009, **20**(3), 240.
31. J. Liu, J. Li and J. Tang, *Food Bioprod. Process.*, 2010, **88**(2–3), 215.
32. A. Alupului and V. Lavric, *Rev. Chim.*, 2009, **60**(5), 497.
33. C. C. Teo, S. N. Tan, J. W. H. Yong, C. S. Hew and E. S. Ong, *J. Sep. Sci.*, 2009, **32**(4), 613.
34. A. Periche, M. L. Castello, A. Heredia and I. Escriche, *Plant Foods Hum. Nutr.*, 2015, **70**(2), 119.
35. E. Yildiz-Ozturk, A. Nalbantsoy, O. Tag and O. Yesil-Celiktas, *Ind. Crops Prod.*, 2015, **77**, 961.
36. J. Šic Žlabur, S. Voća, N. Dobričević, M. Brnčić, F. Dujmić and S. Rimac Brnčić, *Int. Agrophys.*, 2015, **29**, 231.
37. M. Libik-Konieczny, E. Capecka, E. Kąkol, M. Dziurka, A. Grabowska-Joachimiak, E. Sliwinska and L. Pistelli, *Sci. Hortic.*, 2018, **234**, 10.
38. J. Pól, O. B. Varadová, P. Karásek, M. Roth, K. Benešová, P. Kotlaříková and J. Čáslavský, *Anal. Bioanal. Chem.*, 2007, **388**(8), 1847.
39. Y.-H. Choi, I. Kim, K. D. Yoon, S. J. Lee and C. Y. Kim, *Chromatographia*, 2002, **55**, 617.
40. S. K. Yoda, M. O. M. Marques, A. J. Petenate and M. A. A. Meireles, *J. Food Eng.*, 2003, **57**(2), 125.
41. A. Erkucuk, I. H. Akgun and O. Yesil-Celiktas, *J. Supercrit. Fluids*, 2009, **51**(1), 29.
42. F. Ni, J. Ammann and A. Mabud, *J. AOAC Int.*, 2007, **90**(5), 1365.
43. R. Shah, L. S. de Jager and T. H. Begley, *Food Addit. Contam. Part A*, 2012, **29**(12), 1861.
44. B. Shafii, R. Vismeh, R. Beaudry, R. Warner and A. D. Jones, *Anal. Bioanal. Chem.*, 2012, **403**(9), 2683.
45. P. Montoro, I. Molfetta, M. Maldini, L. Ceccarini, S. Piacente, C. Pizza and M. Macchia, *Food Chem.*, 2013, **141**(2), 745.
46. M. Molina-Calle, V. Sanchez de Medina, M. P. Delgado de-la-Torre, F. Priego-Capote and M. D. Luque-de-Castro, *Talanta*, 2016, **154**, 263.
47. S. Soufi, G. D'Urso, C. Pizza, S. Rezgui, T. Bettaieb and P. Montoro, *Food Chem.*, 2016, **190**, 572.
48. A. U. Jackson, A. Tata, C. Wu, R. H. Perry, G. Haas, L. West and R. G. Cooks, *Analyst*, 2009, **134**(5), 867.
49. S. S. Chang and J. M. Cook, *J. Agric. Food Chem.*, 1983, **31**, 409.
50. G. T. Kroyer, *LWT – Food Sci. Technol.*, 1999, **32**, 509.
51. U. Wölwer-Rieck, W. Tomberg and A. Wawrzun, *J. Agric. Food Chem.*, 2010, **58**(23), 12216.
52. Y. Kakigi, T. Suzuki, T. Icho, A. Uyama and N. Mochizuki, *Food Addit. Contam. Part A*, 2013, **30**(12), 2043.

53. A. Tada, K. Ishizuki, J. Iwamura, H. Mikami, Y. Hirao, I. Fujita, T. Yamazaki, H. Akiyama and Y. Kawamura, *Am. J. Anal. Chem.*, 2013, **04**(04), 190.
54. G. E. Morlock, S. Meyer, B. F. Zimmermann and J.-M. Roussel, *J. Chromatogr. A*, 2014, **1350**, 102.
55. H. Sakai, A. Yamashita, M. Tamura, A. Uyama and N. Mochizuki, *Food Addit. Contam. Part A*, 2015, **32**(6), 808.
56. Y.-H. Wang, B. Avula, W. Tang, M. Wang, M. A. Elsohly and I. A. Khan, *Food Addit. Contam. Part A*, 2015, **32**(5), 674.
57. Y. Kitada, M. Sasaki, Y. Yamazoe and H. Nakazawa, *J. Chromatogr. A*, 1989, **474**(2), 447.
58. S. Heeren, P. Schultze, G. Meurer and G. Flick, *VDLUFA-Schriftenr.*, 2010, **66**, 393.
59. E. Jooken, R. Amery, T. Struyf, B. Duquenne, J. Geuns and B. Meesschaert, *J. Agric. Food Chem.*, 2012, **60**(42), 10606.
60. Y.-G. Zhao, X.-H. Chen, S.-S. Yao, S.-D. Pan, X.-P. Li and M.-C. Jin, *Anal. Sci.*, 2012, **28**(10), 967.
61. Y. Hashimoto, M. Moriyasu, S. Nakamura, S. Ishiguro and M. Komuro, *J. Chromatogr. A*, 1978, **161**, 403.
62. H. C. Makapugay, N. P. D. Nanayakkara and A. D. Kinghorn, *J. Chromatogr. A*, 1984, **283**, 390.
63. A. D. Kinghorn, D. D. Soejarto and N. P. D. Nanayakkara, *et al.*, *J. Nat. Prod.*, 1984, **47**, 439.
64. M. J. Ahmed and R. M. Smith, *J. Sep. Sci.*, 2002, **25**, 170.
65. A. S. Dacome, C. C. da Silva, C. E. M. da Costa, J. D. Fontana, J. Adelmann and S. C. da Costa, *Process Biochem.*, 2005, **40**(11), 3587.
66. Tosoh Bioscience, *TSKgel Amide-80*, https://www.separations.us.tosohbioscience.com/solutions/hplc-columns/normal-phase hilic (acccssed April 2018).
67. Phenomenex, *Luna HILIC*, http://www.phenomenex.com/Products/HPLCDetail/luna?culture=de (accessed April 2018).
68. Waters, *BEH HILIC*, http://www.waters.com/waters/promotionDetail.htm?id=10140874&alias=Alias_hilic_CHEMISTRY&locale=de_DE (accessed April 2018).
69. Macherey-Nagel, *Nucleodur HILIC*, http://www.mn-nct.com/HPLC-Start/NUCLEODURphases/NUCLEODURHILIC/tabid/11114/language/en-US/Default.aspx (accessed April 2018).
70. Merck Millipore, *Sequant ZIC cHILIC*, http://www.merckmillipore.com/DE/de/product/SeQuant-ZIC-HILIC-Columns,MM_NF-C136100 (accessed April 2018).
71. L. R. Snyder, J. J. Kirkland and J. W. Dolan, *Introduction to Modern Liquid Chromatography*, Wiley, Hoboken, N.J., 2010.
72. B. F. Zimmermann, U. Woelwer-Rieck and M. Papagiannopoulos, *Food Anal. Methods*, 2012, **5**(2), 266.
73. C. Well, O. Frank and T. Hofmann, *J. Agric. Food Chem.*, 2013, **61**(47), 11312.

74. P. Kubica, J. Namiesnik and A. Wasik, *J. Pharm. Biomed. Anal.*, 2016, **127**, 184.
75. A. E. Abou-Arab, A. A. Abou-Arab and M. F. Abu-Salem, *Afr. J. Food Sci.*, 2010, **4**(5), 269.
76. T. Bililign, J. C. Moore, S. Tan and A. T. Leeks, *J. Agric. Food Chem.*, 2014, **62**(6), 1384.
77. A. M. Hutapea, C. Toskulkao, P. Wilairat and D. Buddhasukh, *J. Liq. Chromatogr. Relat. Technol.*, 1999, **22**(8), 1161.
78. C. Gardana, P. Simonetti, E. Canzi, R. Zanchi and P. Pietta, *J. Agric. Food Chem.*, 2003, **51**(22), 6618.
79. T. Rajasekaran, P. Giridhar and G. A. Ravishankar, *J. Sci. Food Agric.*, 2007, **87**(3), 420.
80. V. S. P. Chaturvedula, I. Prakash and I. Zamora, *Food Nutr. Sci.*, 2014, **05**(17), 1711.
81. USP General chapter (1225), in *USP 36-NF 31*, ed. The United States Pharmacopeial Convention, Rockville, MD, 2012, p. 983.
82. J. M. C. Geuns, *Stevia and Steviol Glycosides*, ed. Euprint, Heverlee, Belgien, 2010.
83. J. Pól, B. Hohnová and T. Hyötyläinen, *J. Chromatogr. A*, 2007, **1150**(1–2), 85.
84. B. F. Zimmermann, *Food Chem.*, 2018, **241**, 150.
85. B. C. Hoekstra, K. A. Chamberlain, J. S. Traub, S. F. Baugh and S. K. Venkataraman, Poster at the 57th International Congress and Annual Meeting of the Society for Medicinal Plant and Natural Product Research, August 16–20, 2009, Geneva, Switzerland; *Planta Med.*, 2009, **75**, 47.
86. J. F. Clos, G. E. DuBois and I. Prakash, *J. Agric. Food Chem.*, 2008, **56**(18), 8507.
87. V. S. P. Chaturvedula and I. Prakash, *Int. J. Res. Phytochem. Pharmacol.*, 2011, **1**(2), 88.
88. M. S. Ahmed and R. H. Dobberstein, *J. Chromatogr. A*, 1982, **245**(3), 373.
89. M. S. Ahmed and R. H. Dobberstein, *J. Chromatogr. A*, 1982, **236**(2), 519.
90. D. H. Hurum, B. M. de Borba and J. S. Rohrer, *Chromatogr. Online*, 2010, (2), 1.
91. D. Cabooter, R. Amery, E. Jooken, B. Meeschaert and G. Desmet, in *Proceedings of the 4th Stevia Symposium 2010*, ed. J. M. C. Geuns, Euprint, Heverlee, Belgium, 2010, p. 83.
92. Q. Fu, Z. Guo, X. Zhang, Y. Liu and X. Liang, *J. Sep. Sci.*, 2012, **35**(14), 1821.
93. M. Koyoma, K. Yoshida, N. Uchibore, I. Wada and K. a. S. T. n. Akjyama, *Shokuhin Eiseigaku Zasshi*, 2005, **46**(3), 72.
94. E. Koyama, *Food Chem. Toxicol.*, 2003, **41**(6), 875.
95. E. Koyama, K. Kitazawa, Y. Ohori, O. Izawa, K. Kakegawa, A. Fujino and M. Ui, *Food Chem. Toxicol.*, 2003, **41**(3), 359.
96. L. Di Donna, F. Mazzotti, I. Santoro and G. Sindona, *J. Mass Spectrom.*, 2017, **52**(5), 290.
97. B. F. Zimmermann, *Rapid Commun. Mass Spectrom.*, 2011, **25**(11), 1575.
98. M. I. Espinoza, J.-P. Vincken, M. Sanders, C. Castro, M. Stieger and E. Agosin, *J. Agric. Food Chem.*, 2014, **62**(49), 11797.

99. K. K. Murray, R. K. Boyd, M. N. Eberlin, G. J. Langley, L. Li and Y. Naito, *Pure Appl. Chem.*, 2013, **85**(7), 1515.
100. A. Richman, A. Swanson, T. Humphrey, R. Chapman, B. McGarvey, R. Pocs and J. Brandle, *Plant J.*, 2005, **41**(1), 56.
101. M. Ohta, S. Sasa, A. Inoue, T. Tamai, I. Fujita, K. Morita and F. Matsuura, *J. Appl. Glycosci.*, 2010, **57**, 199.
102. W. H. Perera, B. Avula, I. A. Khan and J. D. McChesney, *Rapid Commun. Mass Spectrom.*, 2017, **31**, 315.
103. U. Wölwer-Rieck, B. May, C. Lankes and M. Wüst, *J. Agric. Food Chem.*, 2014, **62**(11), 2428.
104. M. Rychlik, in *Fortified Foods with Vitamins: Analytical Concepts to Assure Better and Safer Products*, ed. M. Rychlik, Wiley-VCH, Weinheim, 2011, p. 3.
105. T. Vehovec and A. Obreza, *J. Chromatogr. A*, 2010, **1217**(10), 1549.
106. A. Oehme, M. Wüst and U. Wölwer-Rieck, in *Proceedings of the 7th Stevia Symposium*, ed. J. M. C. Geuns, Euprint, Belgium, Heverlee, 2013, p. 139.
107. J. Metivier and A. M. Viana, *J. Exp. Bot.*, 1979, **30**(4), 805.
108. A. D. Kinghorn, D. D. Soejarto, N. P. I. Nanayakkara, C. M. Compadre, M. C. Macapugay, J. M. Hovanec-Brown, P. J. Medon and S. K. Kamath, *J. Nat. Prod.*, 1984, **47**(3), 439.
109. F. Fullas, J. Kim, C. M. Compadre and A. D. Kinghorn, *J. Chromatogr. A*, 1989, **464**, 213.
110. B. Nikolova-Damyanova, V. Bankova and S. Popov, *Phytochem. Anal.*, 1994, **5**(2), 81.
111. M. Saifi, A. Ali, M. Saini, N. Nasrullah, S. Khan and M. Z. Abdin, *Int. J. Pharm. Pharm. Sci.*, 2014, **6**(5), 465.
112. J. Lui and S. F. Y. Li, *J. Liq. Chromatogr.*, 1995, **18**(9), 1703.
113. P. Mauri, G. Catalano, C. Gardana and P. Pietta, *Electrophoresis*, 1996, **17**(2), 367.
114. B. Ayyappa, S. Kanchi, P. Singh, M. I. Sabela, M. Dovey and K. Bisetty, *J. Iran. Chem. Soc.*, 2015, **12**(1), 127.
115. V. Pavlíček and P. Tůma, *Food Chem.*, 2017, **219**, 193.
116. V. Pieri, A. Belancic, S. Morales and H. Stuppner, *J. Agric. Food Chem.*, 2011, **59**(9), 4378.
117. A. Tada, K. Takahashi, K. Ishizuki, N. Sugimoto, T. Suematsu, K. Arifuku, M. Tahara, T. Akiyama, Y. Ito, T. Yamazaki, H. Akiyama and Y. Kawamura, *Chem. Pharm. Bull.*, 2013, **61**(1), 33.
118. L. K. Hearn and P. P. Subedi, *J. Food Compos. Anal.*, 2009, **22**(2), 165.
119. C. Yu, K. Xu and Y. Shi, *Energy Procedia*, 2011, **5**, 855.
120. Q. Tang, Di. Min, T. Hu, G. Kai and Y. Wang, *Med. Plant*, 2014, **5**(4), 33.
121. P. V. Jentzsch, S. Torrico-Vallejos, S. Mendieta-Brito, L. A. Ramos and V. Ciobotă, *Vib. Spectrosc.*, 2016, **83**, 126.
122. U. Bartholomees, T. Struyf, O. Lauwers, S. Ceunen and J. M. C. Geuns, *Food Chem.*, 2016, **190**, 270.
123. M. Nowotny, *Entwicklung und Bewertung von Verfahren zur Gewinnung der Süßstoffe aus Stevia rebaudiana*, Dissertation, University of Stuttgart-Hohenheim, 1995.

124. J. M. C. Geuns, J. Buyse, A. Vankeirsbilk and E. H. M. Temme, *Exp. Biol. Med.*, 2007, **232**(1), 164.
125. V. J. Y. Minne, F. Compernolle, S. Toppet and J. M. C. Geuns, *J. Agric. Food Chem.*, 2004, **52**(9), 2445.
126. J. P. Wald and G. E. Morlock, *J. Chromatogr. A*, 2017, **1506**, 109.
127. T. Rajasekaran, A. Ramakrishna, K. Udaya Sankar, P. Giridhar and G. A. Ravishankar, *Food Biotechnol.*, 2008, **22**(2), 179.
128. D. Yang and B. Chen, *J. Agric. Food Chem.*, 2009, **57**(8), 3022.
129. R. R. Catharino and L. S. Santos, *Food Chem.*, 2012, **133**(4), 1632.
130. L.-W. Lee, T.-H. Lee, C.-T. Lin, T. Chen and P.-Y. Lin, *Int. J. Spectrosc.*, 2012, 1.
131. V. Jaitak, A. P. Gupta, V. K. Kaul and P. S. Ahuja, *J. Pharm. Biomed. Anal.*, 2008, **47**(4–5), 790.
132. K. Chester, E. T. Tamboli, M. Singh and S. Ahmad, *J. Pharm. BioAllied Sci.*, 2012, **4**(4), 276.
133. S. V. Londhe and S. M. Nanaware, *J. AOAC Int.*, 2013, **96**(1), 24.
134. L. Z. Wang, B. C. Goh, L. Fan and H. S. Lee, *Rapid Commun. Mass Spectrom.*, 2004, **18**(1), 83.
135. S. Purkayastha, G. Pugh, B. Lynch, A. Roberts, D. Kwok and S. M. Tarka, *Regul. Toxicol. Pharmacol.*, 2014, **60**(2), 259.
136. S. Purkayastha, S. Bhusari, G. Pugh, X. Teng, D. Kwok and S. M. Tarka, *Regul. Toxicol. Pharmacol.*, 2015, **72**, 646.
137. A. Roberts, B. Lynch, R. Rogerson, A. Renwick, H. Kern, M. Coffee, N. Cuellar-Kingston, A. Eapen, C. Crincoli, G. Pugh, Jr., S. Bhusari, S. Purkayastha and M. Carakostas, *Regul. Toxicol. Pharmacol.*, 2016, **79**, 91.

CHAPTER 5

Presentation and Analysis of Other Constituents in the Leaves: Polyphenolics in Stevia rebaudiana *Leaves*

N. KUHNERT* AND H. KARAKÖSE

Department of Life Sciences and Chemistry, Jacobs University Bremen, 28759 Bremen, Germany
*E-mail: n.kuhnert@jacobs-university.de

5.1 Introduction

Stevia rebaudiana requires little introduction in this chapter. Steviol glycosides are considered to be the main plant secondary metabolites of interests, adding value to the plant material due to their sweetness.[1] Both in the European Union and the US, purified steviol glycosides have been approved as food additives and found application in many calorie-reduced foods. However, many consumers purchase stevia leaf material on markets or *via* the internet or grow stevia in their own garden using the leaf material directly. Such leaf material is pending approval as a novel food both in the EU and the US. The use of the leaf material will inevitably lead to human consumption of all further secondary metabolites produced by this plant. Many of these

Food Chemistry, Function and Analysis No. 7
Steviol Glycosides: Cultivation, Processing, Analysis and Applications in Food
Edited by Ursula Wölwer-Rieck

Published by the Royal Society of Chemistry, www.rsc.org

metabolites add additional health benefits to the leaf material adding extra value to the product. Additionally, approval of the leaf material will maintain a livelihood for farmers and agricultural cooperatives cultivating the undemanding stevia crop on "poor soils".

5.1.1 Polyphenolics in Stevia

The major class of secondary metabolites in *Stevia rebaudiana* is next to terpenes, polyphenols. Polyphenols are ubiquitous plant secondary metabolites, whose biological function is still unclear.[2,3] They play an important role in the human diet since all dietary plants contain significant amounts of these compounds.[4,5] They are generally considered as beneficial for human health as evidenced by countless epidemiological and clinical studies, reducing the incidence of multifactorial diseases.[6,7] Polyphenolics can be subdivided into several compound classes. Among those, flavonoids and hydroxycinnamates are biosynthesised by *Stevia rebaudiana*.

5.1.1.1 Profiling of Polyphenolic Secondary Metabolites

In 2011 we published a profiling exercise on hydroxycinnamates in stevia.[8] Stevia extracts were analysed by LC-ESI-MS^n in the negative ion mode using an ESI-ion trap mass spectrometer, allowing assignments of compounds to region-isomeric level and also by high-resolution mass spectrometry using ESI-TOF-MS in negative ion mode connected to LC. The latter allowed reliable determination of molecular formulae.

Using the tandem MS based hierarchical scheme for unambiguous identification of quinic acid ester region-isomers introduced by Clifford *et al.*,[9–11] three caffeoyl quinic acids (CQAs) (**1–3**), seven di-CQAs (**4–10**), three FQA (**18, 19, 20**) and one *p*-CoQA (**11**), three CFQAs (**12, 13, 14**), three CSAs (**15, 16, 17**) and four tri-CQAs (**21–24**) were located in the chromatogram and unambiguously identified. All these chemical structures are shown in Figure 5.1. These stevia secondary metabolites can be grouped into the class of chlorogenic acids (CGA), which by definition are hydroxycinnamate esters of quinic acid. The observed CGA profile resembles that of the green coffee bean,[12] yerba mate[13] as well as other members of the *Asteraceae* family.[14]

Next to CGAs a series of flavonoids could be identified in the leaf extracts of stevia. Flavonoids are a class of secondary metabolites that are produced ubiquitously in fruits and vegetables. By definition, flavonoids are compounds with a C6–C3–C6 structure comprising two aromatic rings, one fused as a benzopyran. Flavonoids include several classes of compounds with similar structure having a C6–C3–C6 flavone skeleton (to avoid nomenclature confusions we refer in the text to all compounds with a C6–C3–C6 skeleton as flavonoids). Within different subclasses of flavonoids, further differentiation is based on the number, position and nature of substituent groups attached on the rings. Most flavonoids appear in plants as

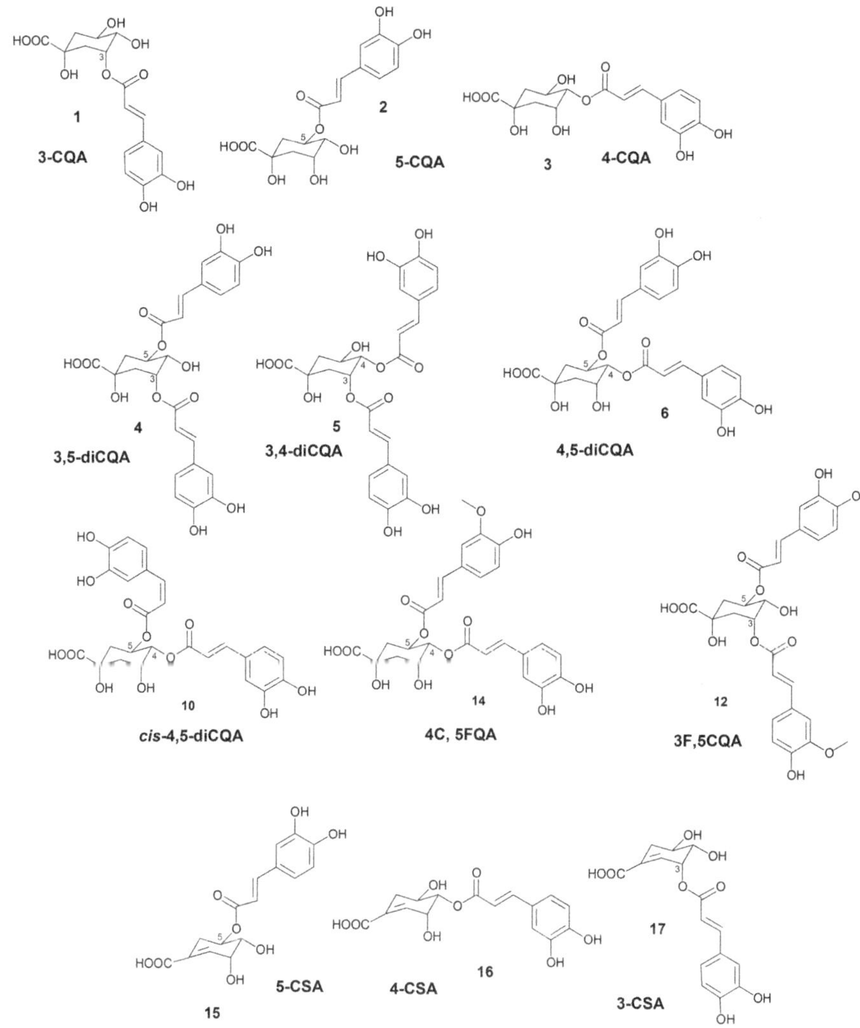

Figure 5.1 Chemical structures of chlorogenic acids **1–17** identified in stevia leaf extracts.

their glycosides **18, 19** with sugars such as glucose, galactose, rhamnose, arabinose, xylose and rutinose conjugated to one or several phenolic OH groups. Flavonoid glycosides have many isomers with the same molecular weight but different aglycone and sugar components at different positions attaching on the aglycone ring.[15]

A total of fifteen peaks in the chromatogram from a stevia leaf extract correspond to flavonoid glycosides with their characteristic fragmentation patterns in tandem MS showing neutral losses of sugar moieties of 162 Da (−$C_6H_{10}O_5$) followed by characteristic fragment spectra of the aglycones quercetin **20** (*m*/*z* 300), kaempferol **23** (*m*/*z* 285), luteolin **22** (*m*/*z* 285) and apigenin

21 (m/z 269) in MS^3. For example, three peaks were located with a m/z value of 447.1 showing, after a neutral loss of 162 Da a base peak at m/z 285 in MS^2. Further fragmentation in MS^3 with the ion at m/z 285.1 as a precursor ion revealed two fragment ions characteristic for kaempferol and one fragment ion characteristic for luteolin. Additional hydrolysis experiments followed by LC-MS analysis confirmed the presence of aglycones of kaempferol, quercetin, luteolin and apigenin by comparison of retention times, high resolution MS data and tandem MS data compared to reference substances in *Stevia rebaudiana* leaves. Therefore, we tentatively assigned the flavonoid glycosides as hexose conjugates of kaempferol, quercetin, luteolin and apigenin **20–23**. Detailed chemical structures of the majority of stevia flavonoids remain open until now. The chemical structures of selected flavonoids and their aglycones are shown in Figure 5.2.

5.1.2 Quantification of Polyphenols in Stevia

Within the framework of the EU-funded project DIVAS (Diversification in Agriculture) *Stevia rebaudiana* was cultivated by agricultural cooperatives in several southern European countries over a period of two years. From these agricultural trials, supplemented by South American and Asian stevia leaf samples, selected polyphenolics were quantified in around 160 samples.[16] The quantification was carried out by LC-UV where possible and by LC-MS in all cases. Quantification was carried out on all three CQAs **1–3**, four diCQAs **4–7** and flavonoid aglycones produced from their glycosides after hydrolysis.

Quantities for selected samples and average values for chlorogenic acids are given in Table 5.1 and for flavonoid aglycones in Table 5.2. All quantities

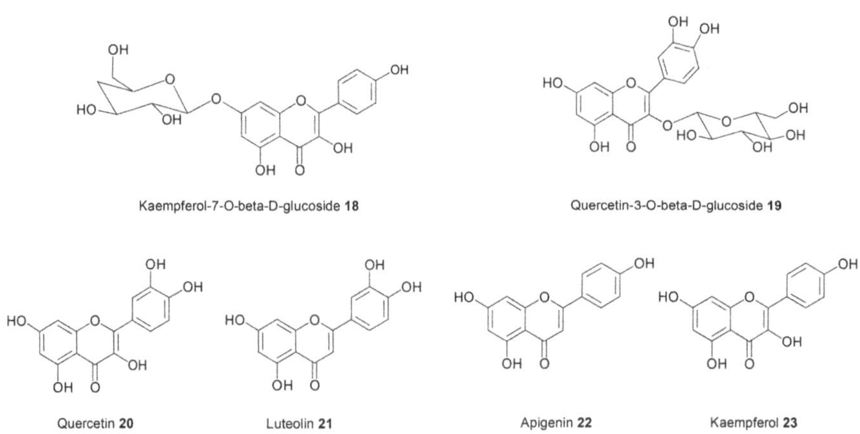

Figure 5.2 Chemical structures of flavonoid glycosides and flavonoid aglycones **18–23** identified in stevia leaf extracts.

Table 5.1 Average values (taken within ± 3 σ) for chlorogenic acids between origins (A–M).

Origin	3-CQA (g per 100 g)	5-CQA (g per 100 g)	4-CQA (g per 100 g)	Total mono CQA (g per 100 g)	3,5-diCQA (g per 100 g)	4,5-diCQA (g per 100 g)	Total diCQA (g per 100 g)
A	0.224	2.015	0.101	2.336	0.909	0.987	1.757
B	0.323	2.060	0.119	2.575	1.373	1.480	2.079
C	0.278	2.078	0.104	2.604	1.233	1.339	2.460
D	0.269	2.450	0.123	2.780	1.351	1.255	2.471
E	0.202	1.580	0.079	1.861	0.555	0.589	1.144
F	0.342	2.208	0.110	2.674	0.898	1.043	1.941
G	0.284	2.446	0.122	2.883	1.435	1.520	2.890
H	0.197	1.855	0.093	2.145	0.686	0.703	1.389
T	0.104	1.837	0.092	2.093	0.727	0.859	1.586
K	0.441	2.646	0.133	3.090	1.302	1.254	2.335
L	0.206	1.363	0.068	1.637	0.746	0.703	1.448
M	0.310	2.525	0.126	2.956	0.993	1.410	2.328

Table 5.2 Values for flavonoids quercetin, kaempferol, luteolin and apigenin determined after hydrolysis of total polyphenol fraction using HCl/MeOH, determined by LC-MS. n.d. indicates that value was outside linear range.

Sample no	Origin	Variety	Harvest	Year	Kaempferol (mg per 100 g)	Quercetin (mg per 100 g)	Luteolin (mg per 100 g)	Apigenin (mg per 100 g)
21	A	1	II	28.09.2010	82.5	75	12	4
32	B	2	I	11.08.2010	n.d.	78	8	3
41	B	3	I	11.08.2010	70	638	9	n.d.
47	D	3	II	10.09.2010	108	463	n.d.	5
61	K	3	II	15.09.2010	90	455	n.d.	n.d.
87	B	7	I	13.07.2011	190	737.5	14	n.d.
89	B	4	I	2011	65	353	n.d.	n.d.
94	B	5	I	13.07.2011	88	417.5	n.d.	n.d.
114	B	6	II	2011	n.d.	445	7	2
115	M	3	I	2011	73	195	8	4
123	C	3	II	2011	73	318	n.d.	n.d.

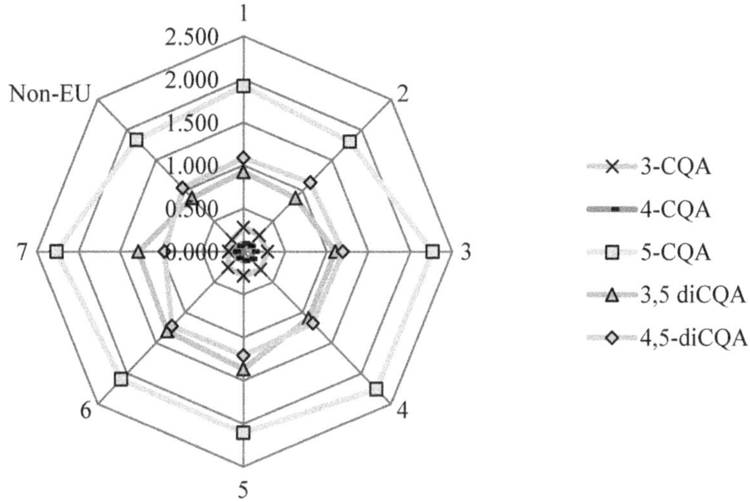

Figure 5.3 Radar plot of individual chlorogenic acid concentrations varying between seven varieties (average values taken within ± 3 σ) and, in comparison, to non-EU samples. Concentrations are given on radial axis in g per 100 g dry leaf material. Outer numbers indicating the seven varieties and non-EU samples; numbers inside the plot are indicating the average concentrations of individual chlorogenic acids.

are given in g per 100 g dried leaf sample. Average quantities for chlorogenic acids are in the range of 3 g per 100 g dry leaves total CGAs. From this data the use of stevia leaf material would significantly contribute to the daily CGA human intake. Stevia leaves from EU origins contain slightly higher CGA concentrations if compared to non-EU samples. An overview on average values is given in Figure 5.3.

It should be mentioned that we have published quantitative data on CGAs in 2011 in our first contribution.[8] The values determined four years later[17] were significantly higher when compared to those published in 2011, probably due to the nature of the leaf material. It must be assumed that older leaf material contains significantly lower CGA concentrations than fresh leaf material.

5.1.3 Further Lessons from Stevia Polyphenolic Analysis

The work within the DIVAS project provided us with a unique and extensive dataset on polyphenolics in an agricultural important economic crop. We obtained the complete polyphenol profile and absolute quantitative data for ten selected polyphenols in more than 150 samples. Never before was such a comprehensive dataset obtained. In addition, the origin, botanical variety date of harvest and agricultural practice used for plant cultivation was meticulously recorded. This unique feature allowed for the correlation of quantitative and qualitative metabolomics data with agricultural practice. Additionally, basic statistical investigations were carried out on the dataset to obtain a fascinating insight into the biochemistry of polyphenols. Some highlights of this work are described in this section.

5.1.4 Plant Polyphenols and UV Exposure

A standard literature hypothesis states that the primary function of polyphenols is to act as UV protection agents in plants.[2,3] Flavonoids show photoprotection between 260–300 nm and hydroxycinnamates between 300–350 nm. Both classes of compounds are present in significant amounts in stevia leaves. This argument is mainly based on genetic studies that have shown that marine plants and organisms do not possess the biosynthetic ability to produce polyphenolic secondary metabolites. Land plants have evolved from green algae that started to colonise land between 630 and 510 million years ago. Shortly after this land colonisation, secondary metabolism of phenolics evolved.

Our data allowed to probe this hypothesis using quantitative data and data on UV exposure of the plants recorded during cultivation or obtained from local weather stations.[17]

In 2008 we reported on the presence of *cis*-caffeoyl derivatives in chlorogenic acids found in plant tissue exposed to UV light.[18] We further showed that UV irradiation induces a photochemical *trans–cis* isomerisation of the caffeoyl moiety (Figure 5.4). Within the DIVAS project we were able to quantify a total of four *cis*-caffeoyl derivatives along with their corresponding *trans*-caffeoyl precursors. The ratio of the two isomers formed shows an excellent linear correlation with UV exposure as measured by sunshine hours per day 30 days prior to the harvest of the leaves. Indirectly, these *cis*-caffeoyl derivatives can serve as a biomarker for sunshine hours of plants and allows measurement of such UV exposure.

However, the sunshine hours could be correlated to all quantitative and semi-quantitative (relative intensities from LC-MS data in the absence of an authentic standard) data for all other phenolics present. This correlation analysis shows (for all flavonoid concentrations) Pearson correlation coefficients below 0.45. For all chlorogenic acids the Pearson correlation coefficients were below 0.5. Consequently, not a single flavonoid and a single chlorogenic acid concentration showed significant correlation with UV exposure. If the initial theory of phenolic photoprotection is correct a significantly better correlation would be expected since a plant would initiate biosynthesis of photoprotection agents following UV exposure. Our experimental data are in clear contradiction to this hypothesis. Further MS imaging data from *Camilla sinensis* leaves have additionally shown that flavonoids show a highly heterogeneous distribution in plant leaf tissue.[19] For a photoprotectant, a homogenous distribution must be expected

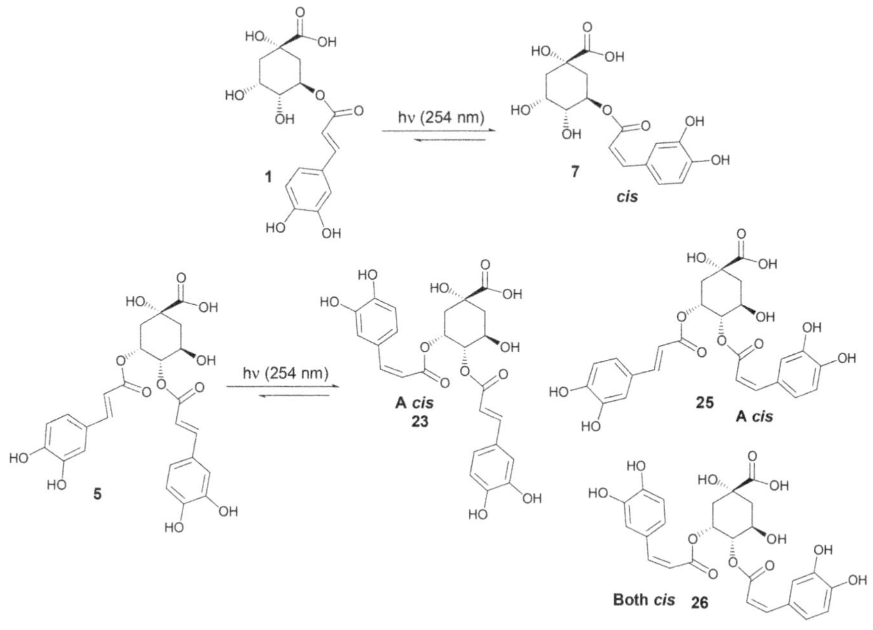

Figure 5.1 Photochemical *cis–trans* isomerisation of chlorogenic acids in stevia leaves upon UV exposure.

(see suntan lotions, which are evenly applied on the skin rather than in patches). Both experimental findings led us to the conclusion that the UV photoprotection theory of polyphenolics is probably wrong. Similar to the antioxidant theory it must be stated to be careful with simple and plausible theories. Phenolics are of course anti-oxidants and they are good absorbers of UV light.[20–22] However, both properties are not at all related to their biological functions but just a nice add-on, which might be beneficial to the plant or not.

As a further important lesson, we demonstrated that UV irradiation, as often used in routine food processing, changes the chemical structure of photoactive food constituents.

5.1.5 Correlation Analysis of Polyphenol Concentrations

With absolute quantitative data for ten polyphenolics available, and further relative quantitative data from LC-MS peak areas for a further 25 compounds, a series of additional correlation analysis could be carried out.[16] A second hypothesis on the biological role of polyphenolics states that polyphenolics are part of the plant defence system against various pests and herbivores including bacteria, viruses, fungi and insects.[7,23] Such plant protection agents should be produced by the plant in higher quantities under a situation of environmental stress. Hence the concentration profile should follow a binomial distribution with two clear maxima.[21] The first maximum corresponds to compound concentrations under normal conditions, whereas the second maximum corresponds to enhanced concentrations of self-defence agents in a situation of stress. Such binomial distributions were for example observed in stevia lipids. Under a situation of temperature stress for example low temperatures the plant produces enhanced quantities of unsaturated fatty acids to keep their lipid bilayers fluid.

Surprisingly, all polyphenol data from our investigations displayed a perfect Gaussian shape concentration distribution. This result was totally surprising and unexpected for us in the light of what was mentioned previously. It is difficult to assign a biological significance to this finding. However, certainly it requires a rethinking and redefinition of the term stress in plant biochemistry and a more thorough investigation of this phenomenon in future large scale agricultural projects.

Additionally, the concentrations of all polyphenolics present in stevia leaves could be correlated in pairs (using the ANOVA method)[17] to probe whether any correlation of two compounds shows a linear relationship resulting in a correlation matrix with around 400 entries. If such a linear relationship exists two explanations are feasible, firstly the two compounds share the same biochemical synthesis pathway and secondly both compounds might be produced following the identical external or internal stimulus.

Indeed, good linear correlations could be observed for all chlorogenic acids as shown in Figure 5.5. This appeared to be expected since this class of compounds share the same biochemical synthesis pathway.

Similarly, most flavonoids and flavonoid glycosides show good linear correlations with R values above 0.7. However, no correlations were observed between classes of compounds in particular no correlations between flavonoid concentrations and chlorogenic acid concentrations. As a conclusion these two classes of compounds are not produced following an identical stimulus such as UV light exposure. These two classes of compounds must play distinct and very different biochemical roles for the plant.

5.1.6 Multivariant Statistical Analysis of Stevia Phytochemical Profiles

With the data in hand we carried out a series of multivariant statistical analysis, in particular using principal component analysis (PCA).[16,24] These types of analysis allow comparison between groups of samples, *e.g.* samples of different origins, different botanical varieties, different harvesting years or harvesting seasons. PCA allows classification of sample groups based on their variance in so called loading plots. Score plots allow identification of unique biomarkers or the identification of all compounds that contribute to the variance between samples.

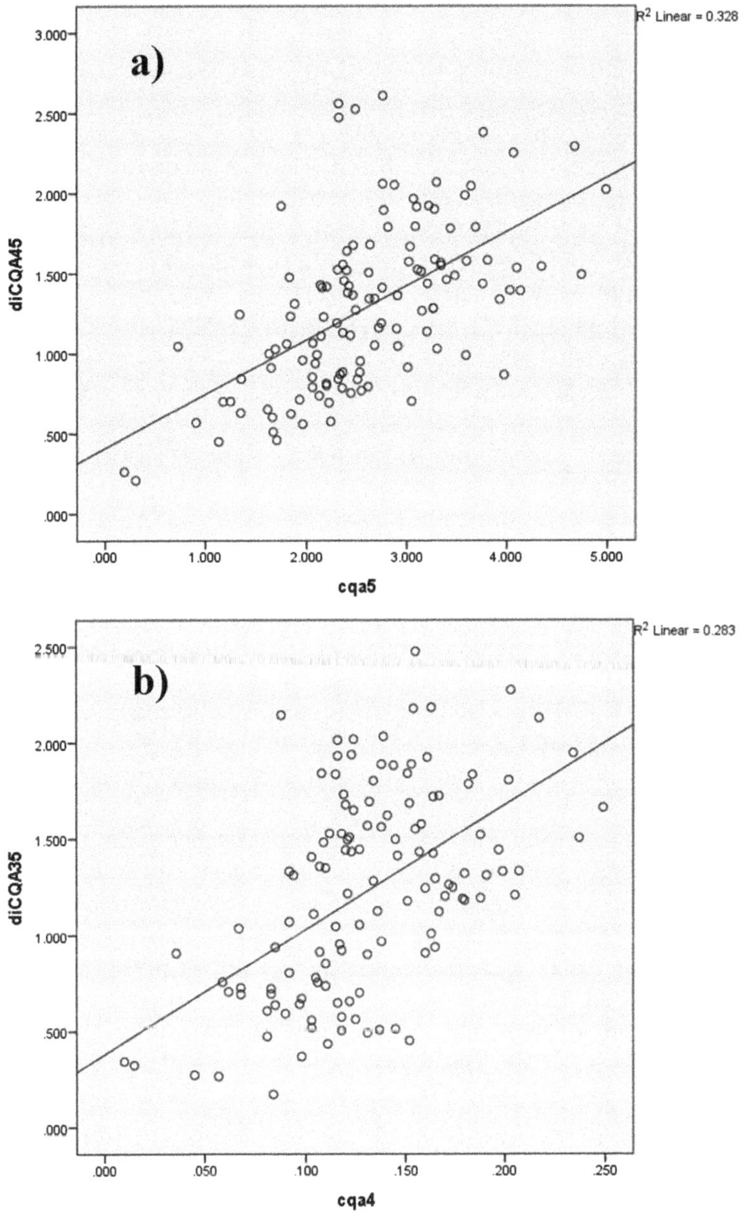

Figure 5.5 Graph showing the correlation between (a) sum of 3,5-diCQA + 4,5-diCQA and 5-CQA and (b) sum of 3,4-diCQA + 4,5-diCQA and 4-CQA.

Acknowledgement

The research leading to these results has received funding from the European Union's Seventh Framework Programme managed by REA-Research Executive Agency http://ec.europa.eu/research/rea(FP7/2007-2013) under grant agreement n° 232250".

References

1. S. K. Goyal, Samsher and R. K. Goyal, *Int. J. Food Sci. Nutr.*, 2010, **61**, 1.
2. E. Haslam, *J. Chem. Ecol.*, 1988, **14**, 1789.
3. E. Haslam, *J. Nat. Prod.*, 1996, **59**, 205.
4. P. Brat, S. George, A. Bellamy, L. Du Chaffaut, A. Scalbert, L. Mennen, N. Arnault and M. J. Amiot, *J. Nutr.*, 2006, **136**, 2368.
5. J. A. Rothwell, M. Urpi-Sarda, M. Boto-Ordonez, R. Llorach, A. Farran-Codina, D. K. Barupal, V. Neveu, C. Manach, C. Andres-Lacueva and A. Scalbert, *Mol. Nutr. Food Res.*, 2016, **60**, 203.
6. A. Scalbert, C. Manach, C. Morand, C. Remesy and L. Jimenez, *Crit. Rev. Food Sci. Nutr.*, 2005, **45**, 287.
7. A. Crozier, I. B. Jaganath and M. N. Clifford, *Nat. Prod. Rep.*, 2009, **26**, 1001.
8. H. Karakoese, R. Jaiswal and N. Kuhnert, *J. Agric. Food Chem.*, 2011, **59**, 10143.
9. M. N. Clifford, K. L. Johnston, S. Knight and N. Kuhnert, *J. Agric. Food Chem.*, 2003, **51**, 2900.
10. M. N. Clifford, S. Knight and N. Kuhnert, *J. Agric. Food Chem.*, 2005, **53**, 3821.
11. M. N. Clifford, S. Knight, B. Surucu and N. Kuhnert, *J. Agric. Food Chem.*, 2006, **54**, 1957.
12. R. Jaiswal, M. A. Patras, P. J. Eravuchira and N. Kuhnert, *J. Agric. Food Chem.*, 2010, **58**, 8722.
13. R. Jaiswal, T. Sovdat, F. Vivan and N. Kuhnert, *J. Agric. Food Chem.*, 2010, **58**, 5471.
14. R. Jaiswal, J. Kiprotich and N. Kuhnert, *Phytochemistry*, 2011, **72**, 781.
15. P. C. H. Hollman and I. C. W. Arts, *J. Sci. Food Agric.*, 2000, **80**, 1081.
16. H. Karakoese, A. Mueller and N. Kuhnert, *J. Agric. Food Chem.*, 2015, **63**, 9188.
17. H. Karakoese, R. Jaiswal, S. Deshpande and N. Kuhnert, *J. Agric. Food Chem.*, 2015, **63**, 3338.
18. M. N. Clifford, J. Kirkpatrick, N. Kuhnert, H. Roozendaal and P. R. Salgado, *Food Chem.*, 2008, **106**, 379.
19. E. Haslam, *Nat. Prod. Rep.*, 1986, **3**, 217.
20. N. Kuhnert, R. Jaiswal, P. Eravuchira, R. M. El-Abassy, B. von der Kammer and A. Materny, *Anal. Methods*, 2011, **3**, 144.
21. M. Miquel, D. James, H. Doover and J. Browse, *PNAS*, 1993, **90**, 6208.
22. P. C. H. Hollman, *Ann. Nutr. Metab.*, 2013, **62**, 149.
23. E. Haslam, *Nat. Prod. Rep.*, 1986, **3**, 217.
24. N. Kuhnert, R. Jaiswal, P. Eravuchira, R. M. El-Abassy, B. von der Kammer and A. Materny, *Anal. Methods*, 2011, **3**, 144.

CHAPTER 6

Presentation and Analysis of Other Constituents in the Leaves: Analysis of Lipids and Volatile Terpenes in Stevia rebaudiana

H. KARAKÖSE, A. GOLON, D. SIRBU AND N. KUHNERT*

Department of Life Sciences and Chemistry, Jacobs University Bremen, 28759 Bremen, Germany
*E-mail: n.kuhnert@jacobs-university.de

6.1 Introduction

Stevia rebaudiana belongs to the Asteraceae family of plants and it is native to Paraguay. Stevia especially is known with its high sweetening property due to *ent*-kaurene glycosides present in the leaves.[1] While there are several studies with steviol glycosides[2] and poylphenolic secondary metabolites[3,4] of stevia in the literature, there is a single study reporting its lipids content[5] and one contribution describing its volatile terpenes.[6] To allow a full chemical characterisation of stevia leaves a thorough profiling of its lipids was carried out. This chapter represents a summary of original research that has not been previously published in peer-reviewed journals.

Food Chemistry, Function and Analysis No. 7
Steviol Glycosides: Cultivation, Processing, Analysis and Applications in Food
Edited by Ursula Wölwer-Rieck

Published by the Royal Society of Chemistry, www.rsc.org

Lipids in plant leaves can be classified according to their chemical structure into triacylglycerides (TAGs), phospholipids and terpenoid structures. Gas chromatographic separation with flame ionisation detection or mass spectrometric (MS) detection is the most widely used method for lipid analysis. Prior to GC analysis, lipids have to be converted into derivatives with lower boiling points, such as methyl esters. Fatty acid methyl ester derivatisation is the most commonly method used for analysis with GC technique otherwise referred to as FAME analysis.[7–9]

In the current study, the lipid and volatile terpene profile of stevia leaf chloroform extracts and steam distillation extract was determined by GC-MS. Quantification and identification of fatty acids were achieved based on the FAME standard mixture. The characterisation of intact TAGs was carried out by LC-MS methods and MALDI-MS.[10,11]

6.2 Analysis of Lipid Fraction

Stevia rebaudiana leaf samples (46 in total), cultivated in seven different places around Europe within the DIVAS EU project[12] comprising five different botanical varieties, were subjected to lipid analysis. The total lipids were determined gravimetrically after extraction with a non-polar solvent. Since hexane extraction provided gravimetric values below 1 wt%, a Soxhlet chloroform extraction was selected and optimised for the purpose of quantitative lipid extraction. Table 6.1 shows total lipid values obtained for 46 representative samples comprising at least three samples from all seven varieties and samples from all origins. Total lipid amounts varied between a minimum of 0.5 and a maximum of 7.2% in the leaves with an average value of 4.3% that coincided with the maximum of a Gaussian distribution.

6.2.1 Analysis of the Fatty Acid Profile

The total lipid fraction was subjected to acid hydrolysis and methyl ester derivatisation followed by GC-MS analysis (FAME method) to identify and quantify the individual fatty acid spectrum of all stevia varieties and of representative samples from all origins.[9] Identification of fatty acid methyl esters was achieved through GC-MS by comparison of retention time and mass spectra with a commercial certified reference compound mixture.[7,8] A representative chromatogram is shown in Figure 6.1.

For the determination of the fatty acid profile, GC-FID was applied using the GC-MS data as a reference point. Quantification of methyl esters was carried out using calibration curves for a total of ten methyl esters. Representative values are shown in Table 6.2. The average ratio of saturated to unsaturated fatty acids in stevia was 1 : 3.8. Among the saturated fatty acids, palmitic acid (C16 : 0) was the predominant acid and stearic acid (C18 : 0) was present with considerable variations of quantities (2–24%). Other saturated fatty acids with shorter or longer chain lengths than sixteen were not identified. In terms of unsaturated fatty acids, palmitoleic acid (16 : 1) was identified next to three co-eluting isomers of (18 : 1), with

Table 6.1 Total lipid values determined gravimetrically in wt% (from 46 samples) (for sample details like origin, variety and harvest please refer to ref. 4).

Origin	Variety	Harvesting	%
A	4	II	3.42
B	6	I	7.71
B	7	I	4.19
A	6	II	5.14
A	3	II	5.45
C	7	I	3.05
C	4	III	4.43
C	3	III	6.67
C	3	II	5.59
A	4	II	0.50
C	4	II	4.32
A	3	II	5.92
C	5	II	4.26
C	6	II	4.15
C	7	II	3.92
D	5	I	1.68
D	3	I	5.48
D	6	I	5.16
D	3	II	7.23
D	6	II	4.90
D	4	II	1.09
D	7	I	2.11
D	4	I	3.42
D	5	II	8.23
E	5	I	3.25
E	6	I	4.23
E	1	I	4.27
C	5	I	3.24
D	3	I	3.37
A	2	II	2.63
F	3	I	3.41
E	4	II	2.71
G	4	II	0.54
E	4	I	2.99
D	1	I	0.89
D	2	I	0.98
D	3	II	6.74
D	2	I	5.11
D	2	II	7.05
D	2	I	2.27
G	2	I	3.14
C	2	II	3.69
C	4	I	2.60
C	4	II	5.13
G	4	I	1.46

oleic acid being the predominant compound. Values stated for (18:1) acids in Table 6.2 represent a sum over all three isomers. A series of further polyunsaturated fatty acids were identified, comprising (16:2), (18:3) and (18:2) acids with variable quantities with linoleic acid (18:2)

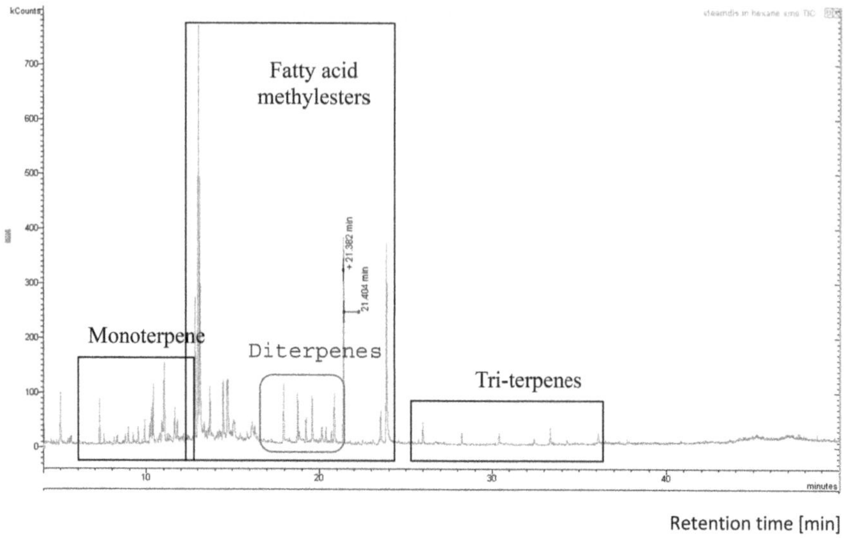

Figure 6.1 GC-MS chromatogram of fatty acid methyl esters obtained from total lipid extracts from *Stevia rebaudiana* leaves from sample Uconor, Var. 4.

being unambiguously identified by comparison to the FAME reference standard.

It should be noted that in plants the fatty acid composition is reported to be a consequence of climatic conditions, in particular temperature, rather than botanical variation.[13] The total lipid content determined is well in line with values from other leafy dietary plants. For example, values for green tea has been determined as 3–5% total lipids, spinach as 4.5% total lipids and the botanically related *Asteraceae* plant lettuce as 4–6% total lipids.[9,14,15] The average fatty acid distribution of stevia leaves is shown in Figure 6.2.

6.2.2 Analysis of Intact Lipids

In a further analysis, we studied the intact lipids, mainly TAGs, in stevia leaves. First, we subjected the crude lipid extracts to MALDI-TOF-MS using HCCA (hydroxyl-cyano coumaric acid) as a matrix. The MALDI-MS spectra were dominated by tri-oleoyl glyceride at *m/z* 885.79 ($C_{57}H_{110}O_6$). No further TAG could be unambiguously identified. For this reason, a second analysis was carried out using LC-ESI-MS on a C-18 reversed phase stationary phase. As the mobile phase, an isocratic elution using ethanol was used with 0.1% NH_4OH added to facilitate TAG ionisation in the positive ion mode.[10] High-resolution mass data were recorded using an ESI-TOF-MS instrument to obtain molecular formulae. Molecular formulae suggestions were accepted for data below 5 ppm mass accuracy. These experiments were followed by tandem-MS measurements using an ion trap instrument to obtain fragment spectra, allowing assignment of individual TAG structure including assignment of

Table 6.2 Quantities of fatty acids in stevia leaves in relative percentages (for sample details like origin, variety and harvest please refer to ref. 4).

			Fatty acid profile (%)						
Origin	Variety	Harvest	C16:2	C16:1	C16:0	C18:3	C18:2	C18:1	C18:0
D	2	II	15.45	9.97	12.04	15.56	10.25	35.09	1.64
D	2	I	20.41	20.52	11.42	2.51	7.57	34.59	2.98
D	2	I	22.03	14.73	9.50	13.49	6.87	30.39	2.99
G	4	II	9.29	1.71	12.26	6.00	11.29	52.06	7.39
G	4	I	3.54	2.10	14.70	8.41	12.44	55.97	2.84
D	1	I	2.09	0.72	14.59	4.70	13.04	57.64	7.23
D	2	II	14.56	9.02	13.01	11.90	10.32	36.11	5.09
D	2	I	4.00	0.51	14.56	5.50	12.61	55.03	7.78
D	3	II	1.21	0.57	12.57	8.35	12.77	48.69	15.83
G	2	I	2.18	1.11	14.86	6.80	11.80	53.53	9.71
G	4	I	5.15	3.38	14.18	9.03	13.75	46.55	7.96
A	2	II	1.49	0.39	11.12	10.81	10.28	49.80	16.11
F	1	I	1.42	0.68	10.35	11.98	10.40	40.32	24.86
B	7	I	7.91	4.77	14.20	6.00	13.59	47.71	5.82
A	6	II	8.44	5.68	15.47	4.10	9.43	52.31	4.57
A	3	II	10.69	6.36	13.41	8.63	9.44	48.57	2.90
C	7	I	0.93	0.46	21.06	5.77	16.42	54.17	1.20
C	3	II	9.05	5.67	16.23	6.39	13.06	43.56	6.03
D	5	I	1.11	0.61	14.03	6.84	10.42	52.11	14.88
D	4	I	8.05	5.38	14.04	4.43	11.29	50.25	6.56
E	5	I	8.27	5.48	15.26	8.71	8.41	45.93	7.94
E	6	I	10.93	7.22	15.29	7.20	9.95	42.27	7.13
E	1	I	7.26	4.38	16.83	4.95	10.70	49.61	6.27
C	5	I	0.63	0.39	15.93	4.80	11.48	60.90	5.86
B	6	I	1.71	0.55	14.79	5.37	11.58	58.93	7.07
D	7	I	2.23	1.31	17.64	6.13	15.83	51.55	5.30
			6.93	4.37	14.21	7.48	11.35	48.22	7.46

all three acyl substituents.[16] This analysis revealed the presence of a further 12 TAG species that included: TG(18:1/18:1/18:1), TG(18:1/18:1/16:0), TG(18:1/18:1/18:0), TG(18:1/18:1/18:2), TG(18:1/16:0/18:2), TG(18:1/18:1/16:2), TG(18:1/16:2/18:2), TG(18:1/18:1/18:3), TG(18:1/16:0/18:3), TG(18:1/18:1/16:1), TG(18:1/18:2/18:0) and TG(18:1/18:3/16:0).

6.2.3 Analysis of Terpenoid Lipophilic Volatiles

Finally, a direct GC-MS analysis of the lipid fraction was carried out supplemented by a GC-MS analysis of a steam distillation extract of *Stevia rebaudiana* leaves. This analysis allowed positive identification of twelve mono and sesquiterpenes using a NIST library search. For the NIST library search, a compound was considered as identified if the database spectrum showed a match with the experimental spectrum with a NIST score of 800 or above. All compounds identified were present in both steam distillation extract and total lipid fraction. Selected structures identified are shown in Figure 6.3.

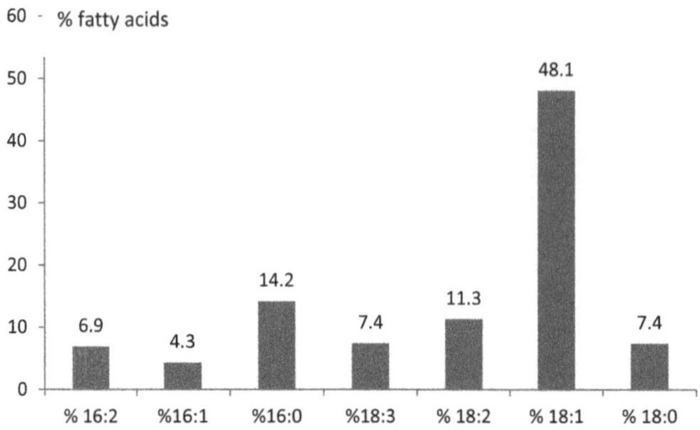

Figure 6.2 Fatty acid profile of average Stevia leaf in %. X:Y denominates the number of carbon atoms in the fatty acid (X) and the number of double bonds in the fatty acid (Y). Data were obtained from 46 samples after lipid hydrolysis, methyl ester derivatisation and GC-MS analysis. A relative RSD of 12.6% was determined.

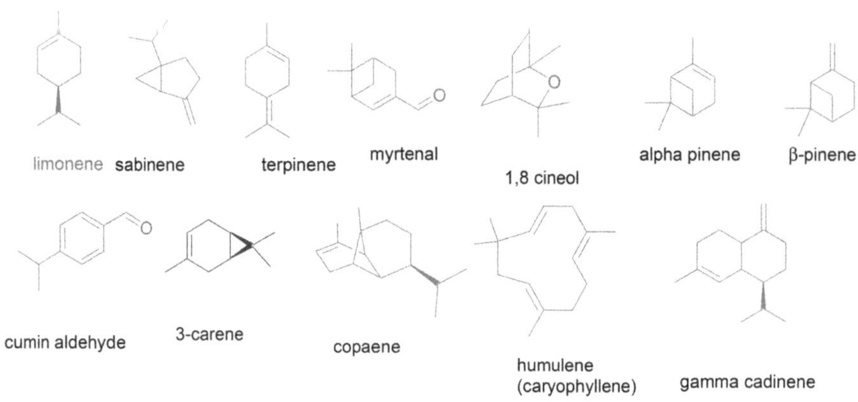

Figure 6.3 Chemical structures of terpenes identified by GC-MS in *Stevia rebaudiana* leaves.

It should be noted that, previously, 25 volatile terpenes have been reported in *Stevia rebaudiana* leaves, but their presence could not be confirmed in this study.[6,16]

In addition to the mono and sesquiterpenes, the GC-MS data show a series of 4–6 triterpenes at longer GC retention times (see Figure 6.1). A NIST search clearly indicates their identity as triterpenes with low NIST scores for steroid and related structures, however, no match in the database allowed compound identification with the exception of the plant hormone gibberellic acid.

Acknowledgements

The research leading to these results has received funding from the European Union's Seventh Framework Programme managed by REA-Research Executive Agency http://ec.europa.eu/research/rea(FP7/2007-2013) under grant agreement n° 232250". We thank Dr Udo Kienle for fruitful and valuable discussions on the science of *Stevia rebaudiana*.

References

1. U. Woelwer-Rieck, *J. Agric. Food Chem.*, 2012, **60**(4), 886.
2. I. Espinoza, J. P. Vincken, M. Sanders, C. Castro, M. Stieger and E. Agosin, *J. Agric. Food Chem.*, 2014, **62**(49), 11797.
3. H. Karakoese, R. Jaiswal and N. Kuhnert, *J. Agric. Food Chem.*, 2011, **59**(18), 10143.
4. H. Karakoese, A. Mueller and N. Kuhnert, *J. Agric. Food Chem.*, 2015, **63**(41), 9188.
5. N. Korobko, Y. Turko, V. Shokun, E. Chernyak, L. Pokrovskii, O. Smetankina, B. Kerimzhanova and U. Baltaev, *Chem. Nat. Compd.*, 2008, **44**, 359.
6. P. L. Cioni, I. Morelli, I. Adolfi, M. Maccia and L. Ceccarini, *J. Essent. Oil Res.*, 2006, **18**, 76.
7. L. Zelles and Q. Bai, *Soil Biol. Biochem.*, 1993, **25**, 495.
8. G. Gutnikov, *J. Chromatogr. B: Biomed. Sci. Appl.*, 1995, **671**, 71.
9. A. Topolewska, K. Czarnowska, L. P. Halinski and P. Stepnowski, *J. Chromatogr. B: Anal. Technol. Biomed. Life Sci.*, 2015, **990**, 150.
10. M. Buchgraber, F. Ullberth, H. Emons and E. Anklam, *Eur. J. Lipid Sci. Technol.*, 2004, **106**(9), 621.
11. P. J. Horn and K. D. Chapman, *Prog. Lipid Res.*, 2014, **54**, 32.
12. DIVAS project coordinated by Dr. Udo Kienle, Universität Hohenheim, Germany; for further details on plant samples and cultivation locations and harvest conditions please see ref. 4.
13. A. Trmolieres, J. P. Dubacq and D. Drapier, *Phytochemistry*, 1982, **21**, 41.
14. C. F. Allen, P. Good and H. F. Davis, *Biochem. Biophys. Res. Commun.*, 1964, **22**, 424.
15. J. W. Drynan, M. N. Clifford, J. Obuchowicz and N. Kuhnert, *Nat. Prod. Rep.*, 2010, **27**(3), 417.
16. A. Martelli, C. Frattini and F. Chialva, *Flavour Fragrance J.*, 1985, **1**, 3.

CHAPTER 7

Antioxidant Capacity of Stevia Leaves

C. BENDER[a,b] AND B. F. ZIMMERMANN*[a,c]

[a]Institut Kurz GmbH, Stöckheimer Weg 1, 50829 Köln, Germany; [b]Istituto Kurz Italia S.r.l., Via Golfo dei Poeti 1/A, 43126 Parma, Italy; [c]Department of Nutritional and Food Sciences, University of Bonn, Endenicher Allee 11–13, 53115 Bonn, Germany

*E-mail: benno.zimmermann@uni-bonn.de

7.1 Definition and Significance of Antioxidant Capacity for Food and Health

7.1.1 Definition

Antioxidant capacity is the ability of a compound or a mixture of compounds, *e.g.* a food, to react as a reducing agent. Such compounds are therefore called antioxidants or (more chemically) reductants. In other words, the antioxidants are oxidised in the presence of oxidising agents (oxidants) and the latter are reduced during this reaction. Among physiological relevant oxidants, reactive oxygen species (ROS) and reactive nitrogen species (RNS) are the most cited. Many of them are characterised by an unpaired electron and are therefore called radicals.

Food Chemistry, Function and Analysis No. 7
Steviol Glycosides: Cultivation, Processing, Analysis and Applications in Food
Edited by Ursula Wölwer-Rieck

Published by the Royal Society of Chemistry, www.rsc.org

7.1.2 Significance of Antioxidant Capacity for Food

Food spoilage can be caused by oxidation. The most common outcomes are fat rancidity, aroma change, discoloration and loss of nutrients (as vitamins). Endogenously present food components and food additives with antioxidant capacity delay oxidation. The latter can be synthetic (such as butylhydroxyanisole (BHA) or butylhydroxytoluene (BHT)), inorganic (such as sulfite) or from a natural source (such as ascorbic acid or tocopherols). Plant extracts containing antioxidants are the subject of intense research. They can satisfy the consumer's demand for "natural" ingredients and can be obtained from by-products of the food production.[1] Moreover, they can have antimicrobial properties. The addition of some plant extracts, *e.g.* from spices, can influence the taste of food and are therefore not always adequate. Popular examples are rosemary and sage extracts containing polyphenols, citrus essential oil containing terpenes and aronia extract containing ascorbic acid.[1–4]

7.1.3 Significance of Antioxidant Capacity for Health

Redox reactions occur constantly in all living organisms as a part of metabolism and catabolism, *e.g.* oxidative phosphorylation, which is the common biochemical pathway of almost all aerobic organisms for releasing energy. The reactions are controlled by the usual biochemical means.

Oxidants are produced by the cells of an organism as part of biochemical reactions or as unintended side products. These compounds can spontaneously react with other functional molecules in the cell like lipids, DNA and amino acids. This can cause severe diseases. During inflammation, the organism produces oxidants in order to neutralise harmful intruders as bacteria or eliminate damaged tissues.[5–7] Additional to the endogenous oxidants, also exogenous oxidants can cause damage to the organism. Exogenous oxidants are consciously introduced to the body (*e.g.* active tobacco smokers, side effect of an X-ray examination) or the body is exposed unintended to the oxidants as pollutants or radiation (*e.g.* sunlight, radioactive radiation). Therefore, the organism possesses several mechanisms to reduce endogenous and exogenous oxidants to harmless molecules. These mechanisms are often rather poetically called "antioxidative defence". Anyhow, oxidative molecules also play an essential role in biochemical signalling.[8–13] Several diseases are known to be linked to an excess of oxidants (so called "oxidative stress"[14–16]) in the cells and even senescence might be caused or accelerated by a decreased ability to reduce oxidants.[17–20]

This led to the simple idea to deliver antioxidants to the body *via* food or food supplements. The antioxidant capacity of single compounds or foods can be measured easily *in vitro*, although a myriad of different analytical methods makes it difficult to compare and judge the results. But the pivotal question is: if a food has an antioxidant capacity as measured *in vitro*, does that increase the antioxidant mechanisms in the body? This effect may be direct or indirect.

7.1.3.1 Direct Antioxidant Effect in the Organism

If an antioxidant present in food should directly react with oxidants in the body, three prerequisites have to be considered: (1) the antioxidants have to be absorbed (bioavailability); (2) the antioxidants have to be transported to the target tissue and (3) the antioxidants have to be present in a sufficient concentration.

The outer layers of the gastro-intestinal tract are in direct contact with the food and its antioxidants, thus they can immediately exert positive effects.[16,21] Another easily accessible organ is the skin. Topic application of antioxidants can abate damage by radiation.[22–24] But any other tissue can be reached by the antioxidants only after absorption *via* the gastro-intestinal tract, for example, the bioavailability of polyphenols is rather low, *i.e.* mostly below 2%. The fate of the undetected polyphenols after ingestion is partly unclear.[25] Most of them are metabolised by the gut microbiota to small molecules as monophenols.[26]

After absorption, most of the polyphenols are metabolised by phase-II enzymes by conjugation with methyl, glucuronide and sulfate moieties. In that way, hydroxyl groups of the molecules are blocked and the antioxidant property of the molecule is decreased or vanished.[27,28] Among the rare cases of polyphenols that circulate as such, *i.e.* not metabolised, in the blood are the galloylated flavonoids like EGCg (epigallocatechin gallate) that occurs in high amounts in green tea.[29]

Given the amount of polyphenols in a food serving, and their low bioavailability, the concentration in the tissues is so low that their reaction with oxidants will not be a significant contribution to the endogenous antioxidant mechanisms.[2–4] Even if a sufficient concentration is reached at the right place in the body, polyphenols are readily eliminated with half times of a few to several hours.[5–7]

7.1.3.2 Indirect Antioxidant Effect in the Organism

Although a direct antioxidant action of the polyphenols is unlikely, there is huge evidence for their beneficial effects. These effects seem to be caused by modulating enzymes, *e.g.* activation of antioxidant enzymes (such as catalase, superoxide dismutase (SOD) or glutathione peroxidase) or inhibition of cellular oxidases, and signal transduction pathways (so that antioxidant mechanisms are activated) and by chelating metal ions that promote the formation of reactive oxidants and by modulating the activity of other enzymes.[8–13] But these modes of action are independent from the antioxidant capacity of the compounds *in vitro*.[14–16]

Given the importance of the right balance of oxidants and antioxidants in the body, this equilibrium is regulated by complex biochemical processes. Therefore, a well-nourished organism will keep this equilibrium regardless of additional antioxidants in food and supplementation might be useless or even noxious.[17–20]

7.2 Determination of Antioxidant Capacity

Generally, the methods for the determination of the antioxidant capacity provoke an oxidation of the antioxidant and the reaction is observed *via* detection of a molecule that is formed or decomposed during the reaction. Often, photometric or fluorescent detection is used. Next, the methods that have most often been used for stevia are presented.

7.2.1 Extraction of the Antioxidants

It is common to all antioxidant assays that the antioxidants must be dissolved in a clear solution. Therefore, prior to the assay, an adequate extraction must be performed. In most cases, mixtures of water and polar organic solvents such as methanol or acetone are used as extraction solvents, because such mixtures dissolve well the polyphenols (an important group of antioxidants) and must be compatible with the following steps of the assay, *i.e.* also the buffers and other chemicals are soluble. But several antioxidants such as fat-soluble vitamins (tocopherols, carotenoids) or large molecules such as condensed tannins or lignans are not soluble or not completely soluble and their contribution to the antioxidant capacity is therefore underestimated or overseen.

7.2.2 Methods for Determination of Antioxidant Capacity

7.2.2.1 Oxygen Radical Absorbance Capacity (ORAC)

Prior to the ORAC measurement, solid samples are extracted with acetone and water, 50 : 50 (a modified extraction for lipophilic antioxidants[16,21,30] is not relevant for stevia samples). Liquid samples must be filtered, if necessary. The sample (extract) is then mixed with fluorescein, a fluorescent molecule that serves as a probe. APPH (2,2′-azobis(2-amidinopropane) dihydrochloride) is added resulting in the formation of peroxyl radicals.[22–24,31] These radicals oxidise the fluorescein to a non-fluorescent molecule, which is observed by measuring the decreasing fluorescence in time intervals of 1 or 2 min. The fluorescence measurement continues until all fluorescein is oxidised, *i.e.* until the fluorescence reaches the baseline (usually 10 to 60 min).[25,31] The antioxidants of the sample delay or slow down the decay of fluorescence. Thus, the area under the fluorescence *vs.* time curve is higher, when a sample with a high antioxidant capacity is present.

Standard solutions of trolox (6-hydroxy-2,5,7,8-tetramethylchromane-2-carboxylic acid) are measured in the same way and the linear relation between the area under the curve and the trolox concentration allows the calculation of a calibration equation.[26,32] The antioxidant capacity is expressed as μmol trolox equivalents (TE) per g or mL of sample (or equivalent units). Due to the high popularity of ORAC results, there is some confusion, when laypersons or journalists cite ORAC values as a number without unit or use the “ORAC unit" (that is not defined) or are not able to convert μmol to mmol, and so on.[27,28,32]

The ORAC protocol was published by Ou *et al.*[33] and has been further automatised by the same working group.[30] As described in these articles, a well-plate reader is essential for an efficient sample handling.

The ORAC assay measures readily the antioxidant capacity of such types of antioxidants that delay the reaction between the radical and fluorescein (visible as a lag time) as well as such types of antioxidants that slow down this reaction (visible as a slower decrease of fluorescence), because the course of the reaction is tracked until the end, *i.e.* complete oxidation of fluorescein.[34] The endpoint, *i.e.* the measuring time, is determined by the end of the reaction and is not arbitrarily as in some other assays.

The ORAC assay is not standardised by an official body, but the protocol of Ou *et al.*[33] is commonly used and hence the results are comparable to each other.

There are only two publications dealing with the ORAC analysis of stevia leaves.[35,36] In the study of Bender *et al.*[36] dried stevia leaves from different origins were compared. They all had an ORAC value of around 1000 μmol TE g^{-1}, with the exception of one sample with 250 μmol TE g^{-1}. The latter value is very similar to that found for the stevia stem sample. In the study of Lemus-Mondaca *et al.*[35] fresh stevia leaves had an ORAC value of 223 μmol TE per 100 g dry matter. This is virtually the same result as for the sample with extraordinary low antioxidant capacity in the aforementioned study. Unexpectedly, the ORAC value increased after drying at 30, 40, 50, 70 and 80 °C, but not at 60 °C.

These results can be compared to ORAC values of other foods. Such values were compiled by the U.S. Department of Agriculture.[37] It is fair to compare the ORAC values of stevia powders only to foods with a very low water content (<10%). The stevia leaf powders have an ORAC value two times higher than dark chocolate or cocoa powder. Similar results were found for dried spices as parsley, sage, rosemary and thyme. Also, sorghum and sumac bran showed high values, *i.e.* two- to four-fold higher values than the stevia samples.

In another study,[38] an aqueous stevia leaf extract was added during the production of a raspberry juice at different stages of the process. The resulting juices had ORAC values between 30 000 and 40 000 μmol TE L^{-1}, while the juice without addition of stevia and the stevia extract alone had only 17 000 and 4000 μmol TE L^{-1}, respectively. This indicates a synergistic effect of the antioxidants from the raspberries and the stevia leaves.

7.2.2.2 DPPH (2,2-Diphenyl-1-picrylhydrazyl)

DPPH is a stable compound with an unpaired electron, thus, a radical. Its colour (absorption at 515 nm) disappears after reaction with an antioxidant (or with another radical).[39] The method suggests itself as follows: the reaction between the antioxidants of a sample with DPPH is observed by absorbance of the solution at 515 nm. Different concentrations of the sample are measured and the concentration that reduces the absorbance by 50% is interpolated: this is the IC_{50} value (also known as EC_{50}) representing the antioxidant capacity. Thus, a low IC_{50} value corresponds to a high antioxidant capacity. Some researchers do not calculate the IC_{50} value, but measure a given concentration of a compound or extract and report the percentage of inhibition, *i.e.* the reduction of the absorbance. Of note, the absorbance should be read, when the reaction is completed, considering that the reaction time is different for different antioxidants and can be longer than 30 min. Nevertheless absorbance is often read after 6 min.[31,40]

The method is not standardised and the extraction protocol is not defined. Only organic solvents, especially alcohols, can be used due to the solubility of DPPH. Therefore, hydrophilic antioxidants are underestimated.[41] Several variants regarding concentration of the chemicals, solvents or temperature have been used and the results are not comparable since the varying conditions lead to different results.[31,40]

By the DPPH assay, some rather consistent IC_{50} values were published for stevia extracts: 9.3 μg mL^{-1} (methanol–water extract),[42,43] and 8.0 μg mL^{-1} (ethanol–water extract),[44] but also 93.5 μg mL^{-1} (ethanol extract),[45] 250 μg mL^{-1} (methanolic extract)[46] and 825 μg mL^{-1} (methanol–water extract).[47] The value of 52.9 mg TE g^{-1} as published by Periche *et al.*[48] is converted to 0.212 μmol TE g^{-1}. Lemus-Mondaca *et al.*[35] published a value of 0.039 μmol TE g^{-1}, which slightly decreased to 0.028 μmol TE g^{-1} after drying. Due to the possible variations of the DPPH assay, the results are not comparable among different publications.

Ahmad *et al.*[49] measured stevia and other plant extracts. They found a slightly higher antioxidant capacity for the methanolic leaf extract of ginkgo than for that of stevia, *i.e.* 82.8% and 77.7%, respectively, while stevia and *Parthenium hysterophorus* showed the same results. Two articles comparing the DPPH values of leaf and callus extracts do not match. According to Kim *et al.*,[50] the leaves have roughly a 1.3 times higher antioxidant capacity than calli (water extracts) and Tadhani *et al.*[51] found a 1.4 times higher antioxidant capacity in calli compared to leaves (water extracts).

7.2.2.3 ABTS (2,2′-Azino-bis(3-ethylbenzothiazoline-6-sulphonic acid)) and TEAC (Trolox Equivalent Antioxidant Capacity)

These two acronyms stand for the same assay, the only difference is, how the results are expressed. The chemical principle is the same as in the DPPH assay: a coloured radical (in this case formed by the reaction of DPPH and persulfate) is bleached by the reaction with the antioxidants of a sample and the decreasing colour (*i.e.* the decreasing absorbance) is measured. The result can be calculated as IC_{50} or by comparing with trolox concentrations; in the latter case the acronym TEAC is used for the assay.

The method is not standardised and the extraction protocol is not defined.

The ABTS assay was applied to stevia samples by Ghanta *et al.*[43] and by Phansawan and Poungbangpho.[52] They found an IC_{50} of 3.0 µg mL^{-1} (methanol–water extract) and around 2.8 µm trolox equivalents per mg regardless of the solvent used for extraction (acetone, methanol, ethanol), respectively. Several variants regarding the concentration of chemicals, solvents or temperature have been used and the results are not comparable since the varying conditions lead to different results.[31]

7.2.2.4 Folin–Ciocalteu

This assay was intended and is often described as a polyphenol determination but, in fact, it reacts with any antioxidant.[53–55] A mixture of sodium tungstate and sodium molybdate (available in commerce as a premixed stock solution) is reduced by the antioxidants of a sample at basic pH and the reduced form is measured photometrically at 760 nm after completion of the reaction (usually 2 h).[56] The absorbance is then compared to that of a series of gallic acid or (epi)catechin standards and the result is expressed as mg gallic acid or (epi)catechin equivalents (equiv.) per g or mL of sample. Sometimes, other polyphenols as quercetin or chlorogenic acid are used as a reference.

The method is not standardised despite of some proposals for standardised protocols[56,57] and the extraction protocol is not defined. Several variants regarding concentration of the chemicals, solvents or temperature have been used and the results are not comparable since the varying conditions lead to different results.[34,56,58]

The published results of total phenolics by the Folin–Ciocalteu assay are compiled in Table 7.1 and range between 25.3 and 65.1 mg gallic acid equiv. g^{-1} in leaves, apart from the markedly higher values by Tavarini and Angelini[46] and the markedly lower result by Lemus-Mondaca *et al.*[35] The results given in mg tannic acid equiv. g^{-1} and mg catechin equiv. g^{-1} cannot be compared to the former. Two articles present results for leaf and callus extracts[44,50]

Table 7.1 Published results of the total phenolics in stevia as measured by the Folin–Ciocalteu assay.

Sample	Extraction solvent	Total phenolics by Folin–Ciocalteu	Source
Leaf	Boiling water	41.5 mg tannic acid equiv. g^{-1}	59
Leaf	Ethanol	61.5 mg gallic acid equiv. g^{-1}	45
Leaf	Ethanol–water	65.1 mg gallic acid equiv. g^{-1}	44
Callus	Ethanol–water	35.9 mg gallic acid equiv. g^{-1}	44
Leaf	Methanol–water	25.3 mg gallic acid equiv. g^{-1}	42
Leaf	Hot water	44.0 mg catechin equiv. g^{-1}	50
Callus	Hot water	131 mg catechin equiv. g^{-1}	50
Leaf	Methanol–water	200–999 mg gallic acid equiv. g^{-1}	46
Leaf	Methanol–water	0.29 mg gallic acid equiv. g^{-1}	35
Leaf	*not clear*	44.4 mg gallic acid equiv. g^{-1}	48

(see Table 7.1), but they are not in agreement as to which vegetable material has the higher antioxidant capacity (this is also true for the DPPH values in the same publications (see above)).

7.2.2.5 FRAP (Ferric Reducing Antioxidant Power)

This assay is based on the reduction of Fe^{3+} to Fe^{2+}. 2,4,6-Tripyridyl-*s*-triazine (TPTZ) forms a complex with both Fe^{3+} and Fe^{2+}, of which the latter absorbs at 595 nm. Absorption is measured after 4 min assuming that the reaction is completed. The reaction mechanism is a pure electron transfer, thus antioxidants reacting by hydrogen atom transfer, *e.g.* glutathione, do not reduce Fe(III). Due to the endpoint at 4 min reaction time, slowly reacting antioxidants are underestimated.[34,58] Tadhani *et al.*[51] found in leaf and callus about 10 mg GAE g^{-1} in both water and methanol extracts, while Najafian *et al.* reported an IC_{50} value of 9233 mg L^{-1} in a methanol–water extract.[47] The FRAP values of Tavarini and Angelini[46] cannot be compared to that due to a different unit: 30 700–133 000 μmol Fe^{2+} g^{-1}.

7.2.3 Bioassays for the Determination of Antioxidant Capacity

Despite the extensive use of different chemical assays to measure the antioxidant capacity of foods, few studies have investigated the antioxidant capacity of foods in a biological environment, and even less studied the antioxidant capacity of stevia leaf. Usually, the bioassays for determination of the antioxidant capacity use the whole cell or its derived biological material (*i.e.* antioxidant enzymes, nucleic acids, *etc.*) treated with the antioxidant alongside with a generator of free radicals, to provoke an oxidation reaction that is measured through the detection of a molecular marker of cellular stress.

However, there are many quantitation procedures in common use with both direct and indirect measurements, including the activity of SOD, catalase, glutathione, the activity of oxidases, the oxidation inhibition using external stressors, among others. As per the *in vitro* evaluation with mere chemical methods, different biological models combined with different bioassays produce different results, making imprudent the comparison of data obtained with different conditions. It should be remembered that the *in vitro* antioxidant potential of a compound provides no information, whether this compound will be absorbed or metabolised *in vivo*. Moreover, as long as the antioxidant effect is indirect, a mere chemical assay does not provide any information, whether this compound will act as an antioxidant in a living organism. To address this point, the bioassays may provide supplementary information before conducting *in vivo* studies, obtaining a fast turnover and cost-effective data on the performance of such antioxidants in a biological model. Nevertheless, how to extrapolate *in vitro* results to an *in vivo* antioxidant effect is another story.

7.2.3.1 Antioxidant Enzymes

Enzymatic systems such as superoxide dismutase, catalase and glutathione peroxidase, act as endogenous antioxidants in cells. It is calculated that in normal healthy conditions, the endogenous antioxidant systems are able to take under control the 99% of the produced ROS.[60] Nevertheless, an imbalance between the production of antioxidants and their elimination contributes to the oxidative damage, which in turn can damage key biomolecules such as proteins, DNA and lipids. But the biological activity of the endogenous enzymatic systems can be modulated by exogenous compounds, thus increasing or decreasing their activities. The increase of the antioxidant enzymes activity as well as the reduction of oxidases activity may contribute to the protection of constituent molecules and thus can be beneficial for cells.

Among the endogenous antioxidant enzymes, SOD and catalase are the most important; SOD catalyses the dismutation of superoxide into hydrogen peroxide or molecular oxygen, while catalase catalyses the decomposition of hydrogen peroxide to water and molecular oxygen. The enzymatic assays measure the consumption of the substrate or the production of the reaction product over time. Colorimetric methods are widely used to indirectly measure the enzymatic activity or its inhibition concentrations, involving a dye substrate that is oxidised upon reaction with one of the molecules implicated in the enzymatic catalysis.

As commented before in this chapter, stevia is a good source of antioxidant compounds, and even when the *in vivo* absorbance of dietary antioxidants is low, they likely modulate the cellular enzymatic system. In this regard, only two studies were conducted with stevia leaves to corroborate its performance at biological levels.

In a comparative study with different plant extracts, researchers detected that the treatment of liver mitochondria isolated from male Wistar rats with stevia leaf extracts do not significantly alter the activity of the mitochondrial SOD, glutathione peroxidase or glutathione reductase.[61] Analogously, a study of the antioxidant potential of stevia water extracts showed that the cytosolic catalase activity is not influenced[36] in the HepG2 human hepatocellular line treated with crude extracts of leaves or stems. The fact that none of the treatments with stevia extracts trigger a significantly change of the antioxidant enzyme system may suggest that the antioxidant activity of stevia is caused by direct reduction of free radicals.

7.2.3.2 Nucleic Acids

Nucleic acid degradation is a damaging process involving intracellular radicals. DNA degradation is probably one of the most damaging processes as DNA repair is a complex mechanism. Severe DNA damage can lead to an ineffective genome repair, thus provoking abnormal transcription of the damaged DNA, cellular death or even uncontrolled cellular division.

The *in vitro* DNA protection assay serves to estimate a preventive action of a sample towards the induced oxidative damage of DNA. Typically, the test involves exposing isolated DNA to an oxidant in the presence of the sample of interest. The DNA damage intensity is compared with an untreated control and migration patterns are visualised on agarose gel subjected to electrophoresis. The intact genomic DNA is visualised as a single band while smearing indicates the degradation of the DNA.

By the DNA protection assay, it was reported that flavonoid-rich extracts from stevia leaves prevent strand scission of DNA.[43] In this work, circular plasmidic DNA was exposed to hydroxyl radicals generated by hydrogen peroxide and an iron(II) catalyst (the so called Fenton reaction), with and without the presence of stevia extract or purified stevioside. Stevia crude extracts neutralised the DNA oxidative damage, while pure stevioside failed to counteract strand scission.

Similarly, cell wall polysaccharide fractions from stevia leaves fully counteracted the human genomic DNA scission by the radical hydroxide.[62]

Interestingly, in the work of Ghanta *et al.* was reported that pure stevioside does not prevent the H_2O_2-induced DNA-scission.

7.2.3.3 Cell-based Antioxidant Capacity

The inhibition of intracellular ROS formation can be measured *in vitro* after the pre-treatment of cells with a sample of interest. The CAA (Cellular Antioxidant Activity) method was first described by Wolfe and Liu.[63] In this method, solid samples are extracted usually in water or PBS (Phosphate Buffer Saline) and diluted in cell culture media. The cytotoxicity of the extracted sample is first measured through common cellular tests (*i.e.* MTT or LDH assays), and then the sample are incubated at not-toxic concentrations with the cells for up to 24 h and with a precursor probe for an hour. The precursor, 2′,7′-dichlorofluorescein diacetate (DCFH-DA), is a molecule that traverses the cell membrane by passive diffusion and once inside the cytoplasm, it is deacetylated by cellular esterases.[64] The intracellular ROS oxidise the deacetylated precursor (DCFH) to highly fluorescent dichlorofluorescein (DCF). Then a ROS generator such as APPH is added to the culture media and passes through the cell membrane. Thus, the oxidation reaction in the cell is accelerated, which is observed by reading the fluorescence formation over time. Samples with antioxidant capacity able to pass through the cellular membrane can delay or block the DCFH oxidation and thus fluorescence does not or more slowly increase compared to a control, *i.e.* without addition of the antioxidant. Of note, contrary to the ORAC assay, the area under the fluorescence curve over time is lower when the sample of interest shows a high intracellular antioxidant capacity.

Usually the fluorescence is measured at 3–5 min intervals for 90 min or until the curves reaches a plateau. The decrease in cellular fluorescence compared to untreated control cells indicates the intracellular antioxidant capacity of the samples.

In the same plate, standard solutions of quercetin are measured, and similar to the ORAC assay, the linear calibration of the area under the fluorescence curve *versus* the standard concentration is calculated. Hence, the cellular antioxidant activity is expressed as μmol quercetin equivalents (QE) per volume or weight of sample.

This method combines several advantages: an inhibition of the oxidation reaction (*i.e.* a lag phase) and a decrease of the reaction speed are both reflected in the result (the same is valid for the ORAC assay); it uses a biological relevant source of radicals (peroxyl radicals); it mimics the reaction between antioxidants and lipids. Furthermore, the CAA considers some key aspects of cell uptake, metabolism, and distribution of the antioxidants within cells.[63] After pre-treatments with the sample of interest and the probe precursor an additional wash step is performed, in order to eliminate the molecules in the cell growth medium that were not absorbed into the cells, so the measurements are indicative of the substances effectively absorbed within the cell. Therefore, the cellular models consider the uptake of antioxidants from the cellular environment and its repartition within the cells. It remains an issue how to extrapolate these data to an *in vivo* situation, for instance, in the case of oral ingestion, it is questionable which and how much of the ingested antioxidants may reach the tissues.

In a study the ORAC values of stevia were compared to those from cell based models using the same samples.[36] After 24 h of cell-treatment (HepG2, a human hepatocarcinoma cellular model) with stevia aqueous extracts obtained from leaves or stems, it was found that the CAA is dose dependent and that leaves gave higher values than stems. CAA values ranged from 61.3 to 14.5 μmol QE g^{-1} except for the stems where it was roughly 8 μmol QE g^{-1}.

Similar results were recently obtained with stevia leaves extract, where a CAA value of 13.6 μmol QE g^{-1} was recorded.[38] The study concerns the influence of stevia addition into antioxidants-rich raspberry juices produced in semi-industrial scale and sweetened with stevia crude extract at different stages of the industrial process. Interestingly, the addition of stevia extracts significantly increased some parameters associated with the antioxidant potential. Interestingly, a synergistic effect was observed: the CAA values of the juices increased with the addition of stevia by about 60%, even when the stevia extract showed much lower CAA and ORAC values regarding juices, as well as lower total phenolic and ascorbic acid contents.

Inversely, it was previously observed that highly purified rebaudioside A or stevioside had no peroxyl radical scavenging capacity as measured by CAA.[36] It is unknown why steviol glycosides showed a low ORAC value but no cellular antioxidant activity in the CAA assay. Likely, there is a link between their structural properties (too large molecules) and the ineffectiveness at a cellular level. Notwithstanding, these findings indicate that there are other stevia constituents that impart the cellular antioxidant activity.

Indeed, following the steviol glycosides the chlorogenic acids (caffeoyl quinic acids, CQA), represent the second most abundant compounds in stevia leaves. Quantities up to 10% w/w have been reported.[36] Although the absorption rate of dietary polyphenols is low *in vivo*, there is enough evidence of the absorption of intact CQA, both *in vitro* and *in vivo* (reviewed by Liang and Kitts[65]), most of the evidence coming from the 5-CQA isomer.

CAA results performed with 5-*O*-CQA indicate that CQA can act as antioxidant at cellular level, being able to scavenge peroxyl radical in the HepG2 liver cells (unpublished results). However, the concentration of CQA necessary to produce a decline in the fluorescence curve is several times higher than the quercetin standard, being roughly 32 µmol quercetin equiv. per mM CQA; even so, this concentration is in line with the CQA content in stevia leaves.

The cellular toxicity has also been used as an indicator of oxidative stress-mediated loss of viable cells. This method measures the cellular response to an external oxidative stressor and its attenuation by pre-treatment with a sample of interest, but this assay is not designed to directly measure the intracellular antioxidant activity. It rather indicates the protective effect against cellular death, which could occur *via* different pathways not always directly connected with oxidation.

The MTT (3-(4,5-dimethylthiazol-2-yl)-2,5-diphenyltetrazolium bromide) bioassay is a widely used colorimetric method to determine, under defined conditions, the number of viable cells. The method principle relies on the cytoplasmic reduction of the yellow-MTT into purple-Formazan only by metabolic active cells. After a solubilising step, the absorbance readings can be recorded by a spectrophotometer.

The percentage of HepG2 cell survival, once pre-treated for 24 h with not-cytotoxic concentrations of stevia ethanolic extract followed by an oxidation treatment with hydrogen peroxide (500 µM for an hour), was recently reported[66] using the MTT assay. Cellular metabolic activity, as a loss of viable cells, was read 24 h after the protection/oxidation treatments. The authors reported a moderate cellular survival compared to control, and a significant survival when the cells were pre-treated with the higher dose (15.62 µg mL^{-1}) of stevia extract.

7.2.3.4 Lipid Peroxidation Inhibition

Another biomarker of antioxidant capacity is the inhibition of the lipid peroxidation. Lipid peroxidation, a catalytic process that is the common consequence of cellular death, can be estimated by the TBARS (thiobarbituric acid-reactive substances) assay as the amount of malondialdehyde thiobarbituric acid (MDA), one of the reaction products in the lipid peroxidation process. Typically, MDA is analysed by HPLC-MS (high-performance liquid chromatography with mass spectrometric detection). The extent of protection is calculated as a decrease in MDA amount and compared with a standard sample.

Two papers reported the lipid peroxidation inhibition exerted by stevia leaf extracts, using liver homogenates isolated from male Wistar rats as a lipid source.[43,62] Positive results were found in both studies indicating an anti-lipoperoxidant capacity of stevia in the TBARS assay. However, divergences in the applied standard (butylated hydroxytoluene or catechin), calculation methods, extraction solvents and partition procedures made a direct comparison impossible. Nevertheless, these results obtained *in vitro* are indicative for the anti-lipoperoxidant potential of stevia leaf extracts and, hence, merit further research.

7.2.3.5 In vivo *Models*

Current *in vitro* data on the antioxidant properties of stevia foster *in vivo* testing. Indeed, the studies cited here mention that stevia leaf extracts have an antioxidant capacity *in vitro*, both with mere chemical tests as well with bioassays. But these data must be further demonstrated *in vivo*.

Most existing *in vivo* studies report the effect of steviolglycosides in animal models, while few data point to the *in vivo* antioxidant capacity of stevia. Only two investigations were reported so far using diabetic rodent models subjected to feeding trials with stevia (leaf powders or polyphenol extract from leaves). In both investigations was demonstrated that stevia-feeding protects the animals against oxidative stress.[67,68] Shivanna *et al.*[67] evaluated the effect of one month feeding with 4% stevia (leaf powder or its polyphenols extract) supplementation to standard diet on diabetic induced Wistar rats. A decrement of SOD and catalase levels in liver homogenates was reported in the group of diabetic rats (streptozotocin induced) without stevia supplementation. These levels were restored to normal levels in the stevia-fed groups. The levels of plasmatic GSH and liver lipid peroxidation (TBARS) were increased in diabetic rats but decreased in the stevia-fed groups.

Similarly, Singh *et al.*[68] investigated the effect of three weeks of daily feeding with a methanolic crude extract from stevia leaves (300 mg kg^{-1} supplementation) on diabetic-induced Swiss Albino mice. A significantly decrease of SOD levels in kidney and pancreas homogenates was reported in the diabetic mice group (alloxan induced) without stevia feeding. SOD level was significantly increased in kidney tissue in stevia-fed diabetic-animals. Also, the levels of GSH and glutathione peroxidase were reduced in liver, kidney and pancreas homogenates from diabetic mice but restored to normal levels in stevia-fed animals. Additionally, the level of TBARS was significantly higher in diabetic animals (in hepatic, renal and pancreatic tissues) but lower in the stevia-fed group regarding the alloxan-uninduced group.

These findings are certainly promising. Several indices of oxidative damage were analysed and the results indicate that the feeding with stevia leaves (or its extracts) would reduce the lipid peroxidation and could restore or even increment the levels of antioxidant enzymes *in vivo*, thus improving the antioxidant status through modulation of the antioxidant enzyme system.

No study was so far reported on the antioxidant effects of stevia crude extracts in humans. However, the supportive evidence so far obtained *in vitro* with chemical tests and bioassays, as well as the results obtained with induced diabetes rodents are an excellent starting point for new *in vivo* research. Stevia antioxidant properties certainly are worthy of further attention.

References

1. A. Schieber, *Annu. Rev. Food Sci. Technol.*, 2017, **8**, 97.
2. B. Halliwell, *Nature*, 2003, **426**, 787.
3. J. Fernández-López, N. Zhi, L. Aleson-Carbonell, J. A. Pérez-Alvarez and V. Kuri, *Meat Sci.*, 2012, **69**, 371.
4. F. Shahidi, *Mol. Nutr. Food Res.*, 2000, **44**, 158.
5. C. Manach and J. Donovan, *Free Radical Res.*, 2004, **38**, 771.
6. A. Santo, H. Zhu and Y. R. Li, *ROS*, 2016, **2**, 245.
7. J. P. E. Spencer, M. M. Abd El Mohsen, A.-M. Minihane and J. C. Mathers, *Br. J. Nutr.*, 2007, **99**, 1.
8. A. Crozier, I. B. Jaganath and M. N. Clifford, *Nat. Prod. Rep.*, 2009, **26**, 1001.
9. W. Dröge, *Physiol. Rev.*, 2002, **82**, 47.
10. R. Rodrigo, A. Miranda and L. Vergara, *Clin. Chim. Acta*, 2011, **412**, 410.
11. Y. M. W. Janssen-Heininger, B. T. Mossman, N. H. Heintz, H. J. Forman, B. Kalyanaraman, T. Finkel, J. S. Stamler, S. G. Rhee and A. van der Vliet, *Free Radical Biol. Med.*, 2008, **45**, 1.
12. H.-S. Kim, M. J. Quon and J.-A. Kim, *Redox Biol.*, 2014, **2**, 187.
13. M. Lorenz, *Am. J. Clin. Nutr.*, 2013, **98**, 1642S.
14. A. Pompella, H. Sies, R. Wacker, F. Brouns, T. Grune, H. K. Biesalski and J. Frank, *Nutrition*, 2014, **30**, 791.
15. H. J. Majima, H. P. Indo, I. Nakanishi, S. Suenaga, K.-I. Matsumoto, H. Matsui, Y. Minamiyama, H. Ichikawa, H.-C. Yen, C. L. Hawkins, M. J. Davies, T. Ozawa and D. K. St Clair, *Arch. Biochem. Biophys.*, 2016, **595**, 54.
16. H. Sies, *J. Nutr.*, 2007, **137**, 1493.
17. B. Halliwell, *Br. J. Clin. Pharmacol.*, 2012, **75**, 637.
18. T. Lu and T. Finkel, *Exp. Cell Res.*, 2008, **314**, 1918.
19. J. W. Finley, A.-N. Kong, K. J. Hintze, E. H. Jeffery, L. L. Ji and X. G. Lei, *J. Agric. Food Chem.*, 2011, **59**, 6837.
20. R. G. Berger, S. Lunkenbein, A. Strohle and A. Hahn, *Crit. Rev. Food Sci. Nutr.*, 2012, **52**, 162.
21. B. Halliwell, *Cardiovasc. Res.*, 2007, **73**, 341.
22. J. A. Nichols and S. K. Katiyar, *Arch. Dermatol. Res.*, 2009, **302**, 71.
23. T. Hatahet, M. Morille, A. Hommoss, J. M. Devoisselle, R. H. Müller and S. Bégu, *Eur. J. Pharm. Biopharm.*, 2016, **108**, 41.
24. P. K. Vayalil, A. Mittal, Y. Hara, C. A. Elmets and S. K. Katiyar, *J. Invest. Dermatol.*, 2004, **122**, 1480.
25. C. Manach, G. Williamson, C. Morand, A. Scalbert and C. Remesy, *Am. J. Clin. Nutr.*, 2005, **81**, 230S.
26. A. Duda-Chodak, T. Tarko, P. Satora and P. Sroka, *Eur. J. Nutr.*, 2015, **54**, 325.
27. A. R. Rechner, G. Kuhnle, H. Hu, A. Roedig-Penman, M. H. van den Braak, K. P. Moore and C. A. Rice-Evans, *Free Radical Res.*, 2009, **36**, 1229.
28. C. Manach, A. Scalbert, C. Morand, C. Remesy and L. Jimenez, *Am. J. Clin. Nutr.*, 2004, **79**, 727.

29. B. F. Zimmermann and M. Gleichenhagen, *Food Chem.*, 2011, **124**, 1543.
30. D. Huang, B. Ou, M. Hampsch-Woodill, J. Flanagan and E. Deemer, *J. Agric. Food Chem.*, 2002, **50**, 1815.
31. K. M. Schaich, X. Tian and J. Xie, *J. Funct. Foods*, 2015, **14**, 111.
32. R. Apak, S. Gorinstein, V. Böhm, K. M. Schaich, M. Özyürek and K. Güçlü, *Pure Appl. Chem.*, 2013, **85**, 957.
33. B. Ou, M. Hampsch-Woodill and R. L. Prior, *J. Agric. Food Chem.*, 2001, **49**, 4619.
34. A. Karadag, B. Ozcelik and S. Saner, *Food Anal. Methods*, 2009, **2**, 41.
35. R. Lemus-Mondaca, K. Ah-Hen, A. Vega-Gálvez, C. Honores and N. O. Moraga, *Plant Foods Hum. Nutr.*, 2016, **71**, 49.
36. C. Bender, S. Graziano and B. F. Zimmermann, *Int. J. Food Sci. Nutr.*, 2015, **66**, 553.
37. D. B. Haytowitz and S. Bhagwat, *U.S. Department of Agriculture, Agricultural Research Service*, 2010.
38. C. Bender, K. V. Killermann, D. Rehmann and H. H. Weidlich, *Beverages*, 2018, **4**(3), 52.
39. W. Brand-Williams, M. E. Cuvelier and C. Berset, *Lebensm. Wiss. Technol.*, 1995, **28**, 25.
40. O. P. Sharma and T. K. Bhat, *Food Chem.*, 2009, **113**, 1202.
41. M. B. Arnao, *Trends Food Sci. Technol.*, 2000, **11**, 419.
42. F. N. Muanda, R. Soulimani, B. Diop and A. Dicko, *Lebensm. Wiss. Technol.*, 2011, **44**, 1865.
43. S. Ghanta, A. Banerjee, A. Poddar and S. Chattopadhyay, *J. Agric. Food Chem.*, 2007, **55**, 10962.
44. I. A. Jahan, M. Mostafa, H. Hossain, I. Nimmi, A. Sattar, A. Alim and S. M. I. Moeiz, *Bangladesh Pharm. J.*, 2010, **13**, 67.
45. S. Shukla, A. Mehta, V. K. Bajpai and S. Shukla, *Food Chem. Toxicol.*, 2009, **47**, 2338.
46. S. Tavarini and L. G. Angelini, *J. Sci. Food Agric.*, 2013, **93**, 2121.
47. S. Najafian, M. Moradi and M. Sepehrimanesh, *Comp. Clin. Pathol.*, 2016, **25**, 743.
48. A. Periche, M. L. Castelló, A. Heredia and I. Escriche, *Food Chem.*, 2015, **172**, 1.
49. N. Ahmad, H. Fazal, B. H. Abbasi and S. Farooq, *Int. J. Phytomed.*, 2010, **2**, 231.
50. I.-S. Kim, M. Yang, O.-H. Lee and S.-N. Kang, *Lebensm. Wiss. Technol.*, 2011, **44**, 1328.
51. M. B. Tadhani, V. H. Patel and R. Subhash, *J. Food Compos. Anal.*, 2007, **20**, 323.
52. B. Phansawan and S. Poungbangpho, *Kasetsart J.: Nat. Sci.*, 2007, **41**, 548.
53. R. J. Robbins, C. Kwik-Uribe, J. F. Hammerstone and H. H. Schmitz, *J. Cardiovasc. Pharmacol.*, 2006, **47**(Suppl 2), S110.
54. D. Huang, B. Ou and R. Prior, *J. Agric. Food Chem.*, 2005, **53**, 1841.
55. J. D. Everette, Q. M. Bryant, A. M. Green, Y. A. Abbey, G. W. Wangila and R. B. Walker, *J. Agric. Food Chem.*, 2010, **58**, 8139.

56. V. Singleton, R. Orthofer and R. Lamuela-Raventos, *Method. Enzymol.*, 1999, **299**, 152.
57. E. A. Ainsworth and K. M. Gillespie, *Nat. Protoc.*, 2007, **2**, 875.
58. R. Prior, X. Wu and K. Schaich, *J. Agric. Food Chem.*, 2005, **53**, 4290.
59. R. Kaushik, P. Narayanan, V. Vasudevan, G. Muthukumaran and A. Usha, *J. Food Sci. Technol.*, 2010, **47**, 27.
60. D. Fusco, G. Colloca, M. R. Lo Monaco and M. Cesari, *Clin. Interventions Aging*, 2007, **2**, 377.
61. L. Vaško, J. Vašková, A. Fejerčáková, G. Mojžišová and J. Poráčová, *In Vitro Cell. Dev. Biol. Anim.*, 2014, **50**, 614.
62. M. K. Francine, W. A. Louise, F. S. Pythagore, A.-T. Barbara, S. Gustave and T. Boudjeko, *J. Coastal Life Med.*, 2014, **2**, 962.
63. K. L. Wolfe and R. H. Liu, *J. Agric. Food Chem.*, 2007, **55**, 8896.
64. C. P. LeBel, H. Ischiropoulos and S. C. Bondy, *Chem. Res. Toxicol.*, 1992, **5**, 227.
65. N. Liang and D. D. Kitts, *Nutrients*, 2016, **8**, 16.
66. V. López, S. Pérez, A. Vinuesa, C. Zorzetto and O. Abian, *Food Funct.*, 2016, **7**, 2107.
67. N. Shivanna, M. Naika, F. Khanum and V. K. Kaul, *J. Diabetes Complications*, 2013, **27**, 103.
68. S. Singh, V. Garg and D. Yadav, *Int. J. Pharm. Pharm. Sci.*, 2013, **5**, 297.

CHAPTER 8

Stevia rebaudiana *Bertoni: Beyond Its Use as a Sweetener. Pharmacological and Toxicological Profile of Steviol Glycosides of* Stevia rebaudiana *Bertoni*

LARA TESTAI*[a,b] AND VINCENZO CALDERONE[a,b]

[a]Department of Pharmacy, University of Pisa, via Bonanno, 6 56120 Pisa, Italy; [b]Interdepartmental Research Center "Nutraceuticals and Food for Health", University of Pisa, via Del Borghetto, 15 56120 Pisa, Italy
*E-mail: lara.testai@unipi.it

8.1 Introduction

Stevia leaves extract is a non-nutritive sweetener derived from the plant *Stevia rebaudiana* Bertoni, which may contain more sweet-tasting compounds, named steviol glycosides (SGs).[35] There are more than 10 SGs present in the leaves, such as stevioside, rebaudioside A, B, C, D, E, F, steviolbioside, rubusoside and dulcoside. The best-characterised are stevioside and rebaudioside A.

Food Chemistry, Function and Analysis No. 7
Steviol Glycosides: Cultivation, Processing, Analysis and Applications in Food
Edited by Ursula Wölwer-Rieck

Published by the Royal Society of Chemistry, www.rsc.org

Stevia leaves were first used by the indigenous people in South America, where it is known as *honey leaf*, as both a food sweetener and as medicine.[17] Centuries later, stevia was introduced (for the same uses) in Japan in the 1970s,[4] the first country to market it.

To date, SGs are widely used as food additives in Europe and in several other countries around the world (such as Canada, USA and Australia) as a substitute for sucrose in many food categories and is named E960.

Stevioside and rebaudioside A are thermally stable under the conditions used in food processing, up to 80 °C for 8 h. SGs are also stable in acidic conditions (pH = 2–4).[40]

8.2 Regulatory Committee Position

Regulatory agencies carried out appropriate toxicokinetic studies in order to know the efficacy and safety profile of SGs.

Several assessments have been made for stevioside as a sweetener by: (1) the Joint FAO/WHO Expert Committee on Food Additives (JECFA);[39,40] (2) the Scientific Committee for Food (SCF);[59–61] (3) Food Standard Australia New Zealand (FSANZ) and also (4) EFSA.[13,14]

By virtue of these considerations, these regulatory agencies authorised the use of SGs and established an acceptable daily intake (ADI) of 4 mg kg^{-1} d^{-1}.

SGs are currently authorised as food additives in the EU for use in several food categories. The final product should contain not less than 95% of the 11 named SGs: stevioside, rebaudioside A, B, C, D, E, F, M, steviolbioside, rubusoside and dulcoside (EFSA).[13–15]

8.3 Low Calorie Sweeteners

SGs are endowed with a taste that is 40 to 300 times sweeter than sucrose; in particular rebaudioside A seems to have the best profile, having a reduced liquorice/metal taste and prolonged sweet sensation when compared with the other SGs.[43]

The sweet taste is a common characteristic of all sweeteners, among which it is possible to distinguish two categories: high-calorie sweeteners and low-calorie sweeteners.

Apart from their sweet taste, sucrose and fructose provide a high caloric intake; potentially being responsible for weight gain and hyperglycaemia. However, some sweeteners have the advantage to be at low or no-calories, such as SGs and potassium acesulfame, aspartame, saccharin and sucralose. Potassium acesulfame is not metabolised and is rapidly eliminated mainly in the urine. Unfortunately, it can pass across the placenta and be excreted in the milk of lactating women and animals. Aspartame has a caloric amount similar to sucrose but considering its higher sweet taste the necessary dose of aspartame leads to virtually cancelling the calories. Aspartame is extensively metabolised, raising phenylalanine concentration, particularly in individuals with phenylketonuria, in which it should not be used. Saccharin is adsorbed but not metabolised and it rapidly eliminated in the urine. Finally, sucralose, structurally similar to sucrose, has been approved in 1989. It is minimally adsorbed and is not metabolised, does not modify blood glucose and it is used for foods and beverages appropriate for diabetics.

8.4 The Pharmocokinetic Profile of SGs

SGs require bioactivation by intestinal bacteria in order to ensure their conversion to steviol, the active metabolite. Indeed, enzymes and acids present in the stomach do not hydrolyse SGs. As reported by Hutapea *et al.*, salivary and also pancreatic amylases, gastric secretions and other enzymes typical of the upper gastrointestinal tract do not change stevioside or rebaudioside A and D.[31,51]

Stevioside and rebaudioside A are converted to steviol by caecal, colonic and faecal bacteria through, probably, *Bacterioides* species.[19] Several studies have been carried out in *in vivo* animal models and overall, they demonstrated that an oral administration of steviol results in a rapid rise of steviol into the portal plasma. However, after the administration of SGs, the steviol detection in portal plasma is much slower and its presence is sustained over a period of hours.[36] Such evidence is probably due to slow metabolisation by colonic bacteria and then to a long and slow increase in portal and systemic levels.

In 2008 a study on Sprague-Dawley rats demonstrated that, after the oral administration of ^{14}C-labelled SGs, the plasma peak was reached between two and eight hours after dosing, and the main metabolite was steviol. The main metabolite of steviol is a glucuronide derivative, produced in the liver and transported *via* bile back to the intestinal tract, where it can be again metabolised (Figure 8.1).[58]

Here, steviol metabolites are excreted predominantly in faeces and only less than 2% in urine 96 h after dosing.[58] Similar results have been reported in pharmacokinetic studies in other animal species for stevioside, rebaudioside A, D and other SGs.[20,21,51] Finally, in humans, Wheeler and colleagues confirmed the pharmacokinetic profile, observing that the peak plasma time was 8 h after administration of stevioside and 12 h after rebaudioside A.[80] The half-life was about 14 h. In humans, the excretion of steviol glucuronide is almost exclusively in urine. This is the major difference between the two species: probably, due to molecular weight, as in rats this value is over the thresholds for biliary excretion.[5]

Finally, in children, SGs are metabolised to glucuronide derivatives and excrete at rates similar to adults.[11]

8.5 The Pharmacodynamic Profile of SGs

A number of animal studies and clinical trials have been performed to investigate the effects of SGs, mostly of stevioside and rebaudioside A, observing that beside the sweet taste, these natural derivatives may offer a multitude of

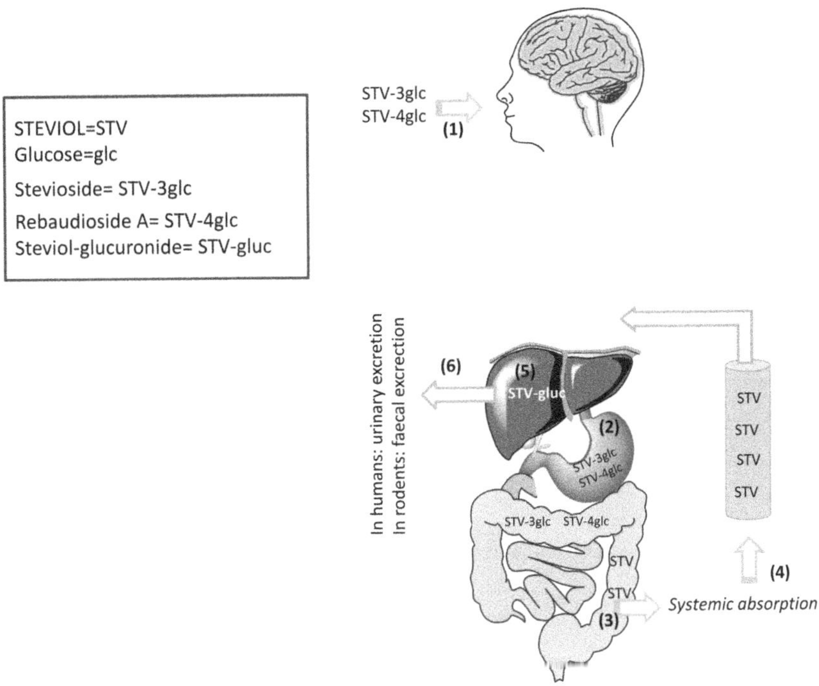

Figure 8.1 Schematic representation of pharmacokinetic profile of SGs, stevioside (STV-3glc) and rebaudioside A (STV-4glc). Usually, SGs are orally taken (1), they are not absorbed in the stomach or upper part of the intestine (2). They are hydrolysed into steviol (STV) in the colon by microbiota enzymes (3) and are then absorbed by systemic circulation (4). STV is metabolised in the liver through the conjugation with glucuronic acid (STV-gluc) (5) and is excreted in the urine of humans and in the faeces of rodents (6).

beneficial effects on health. Indeed, growing evidence points to a nutraceutical use of *Stevia rebaudiana* Bertoni and its SGs.

8.6 Cardiovascular Benefits

A number of studies have demonstrated benefits on the cardiovascular system for stevia extract and for SGs, in particular stevioside and rebaudioside A, whose effects are mediated by the aglycone steviol. Recently, a steviol-analogue diterpene, isosteviol, received great interest for its cardioprotective profile (Figure 8.2).

8.6.1 Antidiabetic Activity

On pancreatic isles isolated from mice, stevioside and its active aglycone, steviol, dose-dependently enhanced the insulin secretion in the presence of high levels of glucose (16.7 mmol L^{-1}). However, at lower concentrations

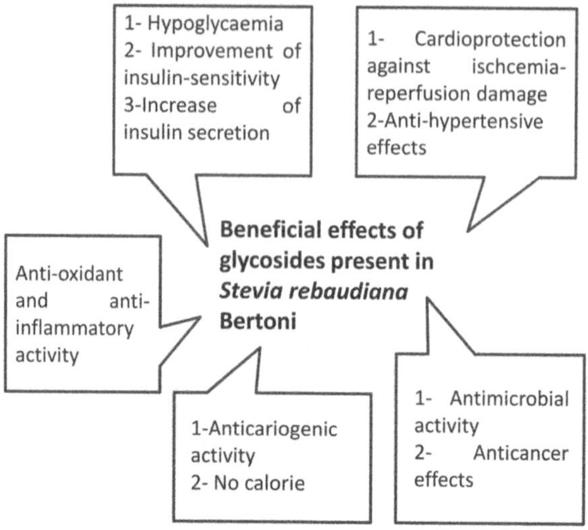

Figure 8.2 Beneficial effects obtained with SGs present in the *Stevia rebaudiana* Bertoni leaves.

of glucose, steviol and stevioside did not significantly influence insulin secretion. Such an insulinotropic effect seems to result from direct action on β cells, leading to the hypothesis, for the first time in 2000, a potential role of these compounds as anti-hyperglycaemic agents in the treatment of Type 2 diabetes mellitus.[41] In 2004, the hypoglycaemic effects of stevioside (250 mg kg^{-1}) and *Stevia rebaudiana* extract (4660 mg kg^{-1}) were evaluated *in vivo* on diabetic rats treated for 8 weeks. In normal rats both stevioside and *Stevia rebaudiana* extract were ineffective. In diabetic rats they improved the serum insulin.[62] Interestingly, Dutta and colleagues carried out a short-term study where the hypoglycaemic effect of stevia (10 mL kg^{-1}) was compared with the conventional drug metformin, and both treatments showed a reduction of blood glucose levels in diabetic rats.[12]

With regards to clinical trials, Curi *et al.*, in 1986, observed in 16 healthy volunteers that the aqueous extract of *Stevia rebaudiana* significantly decreased plasma glucose levels, speculating that stevia extract could increase the mitochondrial respiration rate and inhibit the gluconeogenesis pathway.[9] Later, Mishra tested stevia extract (0.5–1.0 g) on 15 diabetic patients, that used it in place of sugar in beverages for 45 d. A reduction in both postprandial and fasting glucose levels was observed.[49]

Some authors suggested that stevioside can increase insulin secretion and its utilisation, yet SGs are also able to affect the glucose transporter, improving the glucose uptake into the cells.[57] In this regard, Rizzo *et al.* observed that SGs could modulate glucose transport in different types of cells (SH-SY5Y neuroblastoma and HL-60 myeloid leukaemia human cells), similarly to insulin. Indeed, co-treatment with insulin and stevia extract caused a rise of glucose transport, significantly higher than insulin alone; suggesting the existence of a synergism between these substances. Rizzo and colleagues also observed the involvement of the PI3K/Akt pathway. Indeed, the stimulation of insulin secretion through the activation of its own receptor produces phosphorylation of tyrosine receptor kinases and activation of the downstream signal transduction pathway, involving PI3K/Akt. Immunoblots showed an increase of phosphorylated forms of both kinases following the treatment with stevia extract.[57]

Another pathway involved in the antidiabetic action of stevia extract is the PPAR γ receptor. In fact, its expression markedly diminishes in diabetic individuals as well as in animals fed a high fat diet. Assaei and colleagues showed that an administration of stevia extract (400 mg kg^{-1}) for 28 d improved the PPAR γ expression in diabetic rats, similarly to pioglitazone (10 mg kg^{-1}), preserving also the histological characteristics of β cells.[2] Very recently, an aqueous stevioside solution administered in mice at a low-dose (20 mg kg^{-1}) prevented the increase in glycaemia, in an oral glucose tolerance test in alloxan-induced hyperglycaemia. This beneficial effect seems to be due to a preservation of the normal cyto-architectonic arrangement of β islets, paving the way for a nutraceutical use of *Stevia rebaudiana* Bertoni at sweetness doses.[42]

Furthermore, a common clinical condition of subjects affected by Type 2 diabetes is overweightness and obesity. In 2009, two researchers demonstrated that stevia extract could significantly improve serum and liver triglycerides, through upregulation of hepatic PPAR α and then an increase in the expression of PPAR target genes. The authors hypothesised that such an effect could be related with weight loss, but at present no study has been done.[54]

Although a large amount of evidence of the antidiabetic effect has been obtained on the main SGs, stevioside and rebaudioside A, recently also minor SGs have been tested for their anti-hyperglycaemic activity. After a chronic treatment (28 d) with dulcoside A, rebaudioside B, C, D or steviolbioside (20 mg kg^{-1}) in normal or diabetic rats, the glucose tolerance test highlighted the absence of effects on blood glucose.[1]

8.6.2 Antihypertensive Activity

A frequent pathological condition associated with diabetes mellitus is hypertension. In animal models, *Stevia rebaudiana* Bertoni has been shown to reduce blood pressure.[44] Indeed, the intraperitoneal injection of stevioside (25 mg kg^{-1}) had antihypertensive effects and in vasopressin-contracted aortic rings stevioside produced a vasodilation both in the presence and absence of endothelium. Moreover, the intravenous injection of stevioside reduced the blood pressure of spontaneously hypertensive rats.[7]

Besides preclinical evidence, the antihypertensive effects have been reported also in humans. Chan *et al.* carried out a year-long double-blind controlled study on 106 hypertensive subjects assuming daily 750 mg of stevioside or placebo. After three and nine months the stevioside-group exhibited a significant decrease in systolic and diastolic blood pressure.[8] Similar results have been reported by Hsieh *et al.* (2003),[29] in hypertensive patients treated with 1500 mg of stevioside and by Sharma *et al.* (2009) in hypercholesterolemic women, consuming 165 mg mL^{-1} of stevia extract.[63]

8.6.3 Cardioprotective Activity

Cardioprotection has been reported for the diterpene isosteviol, a steviol analogue present in the leaves of *Stevia rebaudiana* Bertoni.

Several studies indicate that isosteviol possesses a variety of biological activities typical for SGs, including anti-hypertensive,[6,30] anti-hyperglycaemic,[46] antioxidant, anti-inflammatory, and antitumor effects.[17,50] In addition, isosteviol relieves I/R injury in rat brains and in rodent hearts.[76–78] In particular, Xu and colleagues have demonstrated for the first time that isosteviol produces cardioprotective effects in rats submitted to myocardial I/R injury, in fact an intravenous administration of the diterpene protected the myocardium against a coronary occlusion reducing the myocardial infarct size, improving cardiac performance and stabilising the electrophysiological properties of the heart. The authors suggested that at least some of the observed beneficial effects of isosteviol could be attributed to stimulation of the mitochondrial ATP-sensitive potassium (mitoKATP) channels.[76] Similar conclusions were later reported by Xu and colleagues in an *ex vivo* model. The involvement of the mitoKATP channel was also hypothesised since a selective blocker, 5-hydroxydecanoic acid, antagonised its protective effects.[77]

Recently, Fan and colleagues observed that isosteviol potentiated the diazoxide-mediated oxidation of flavoproteins in mitochondria, confirming the engagement of mitoKATP channels.[16,45]

Based on these considerations, an exciting strategy could be an effective delivery of isosteviol into the mitochondria to improve its pharmacokinetic profile and, consequently, its pharmacological effects.

With this prospective, an innovative series of isosteviol derivatives, obtained through linkage between the carboxylic group and the cationic triphenylphosphonium (triPP) moiety, have been synthesised.[66] Testai *et al.* demonstrated that these derivatives are endowed with mitochondriotropic properties in rat cardiac mitochondria and in H9c2 cells; moreover such isosteviol derivatives were able to confer a myocardial protection against ischemia-reperfusion at a lower dose than isosteviol.[69]

8.7 Non-cardiovascular Benefits of SGs

Although the cardiovascular properties of SGs, and among these the antidiabetic ones, are the best characterised, other beneficial properties are displayed by stevia derivatives.

8.7.1 Anticariogenic Activity

Dental caries is the most common chronic disease in children. Fortunately, a significant improvement in dental caries worldwide has been attained by the implementation of preventive strategies, such as the use of anticaries natural active compounds, reduction of cariogenic microflora and re-mineralisation of dental hard tissue. On the basis of dental caries there is the consumption of sucrose-based sweeteners; therefore, research of alternative non-cariogenic sweeteners for use in the food industry is a field pursued. Aspartame, saccharin, cyclamate, xylitol and mannitol are valid substitutes of sugar, devoid of the cariogenic properties of sucrose; however, these artificial sweeteners cause weight gain and adverse effects among which the increased risk of tumour. Indeed, an ideal sweetener should be colour-free, non-cariogenic, non-mutagenic, non-heat degradable and obviously not expensive.[28]

Almost all caries are positive *versus Streptomyces mutans* and *Lactobacillus acidophilus*, moreover oral bacteria have been associated with a low pH plaque and it is responsible for the demineralisation of dental tissue.

Several evidences pointed to anticariogenic activity of SGs. Particularly, Gamboa and Chares demonstrated antibacterial activity of five extracts obtained from dried leaves of *Stevia rebaudiana* Bertoni. The bacteria most sensitive to stevia extract were Lactobacilli.[18] In *in vivo* studies, a 10% solution of *Stevia rebaudiana* extract reduced the accumulation of dental plaque between 57 and 82%, if compared with sucrose solution.[10]

Later, three 10% solutions of stevioside, rebaudioside A and sucrose were prepared and tested on *in vitro* biofilm formation and *in vivo* pH of plaque. Using a sucrose solution, a higher *in vitro* biofilm formation and lower pH plaque were observed if compared with *Stevia rebaudiana* Bertoni derivatives,[3] and similar results were obtained in 200 children using toothpaste containing stevioside and rebaudioside A.[82]

8.7.2 Antimicrobial Activity

A large part of our knowledge about the antimicrobial activity of *Stevia rebaudiana* extract is derived from Indian research and by the use of ethanolic extract of leaves,[70] containing the highest concentration of phenols and tannins (if compared with other types of extracts). Preliminary experiments on the crude extract demonstrated an antibacterial activity against several microorganisms, including *Pseudomonas fluorescens*, *Proteus vulgaris*, *Klebsiella pneumonia* and *Streptococcus pneumonia*.[55]

Khan and colleagues also reported that ethanolic and methanolic, chloroformic and ethyl acetate extracts from leaves were the most effective against *Streptococcus aureus* and *Pseudomonas aeruginosa*.[32] Finally, Gosh *et al.* reported on high antimicrobial activity also for an aqueous extract against bacteria and fungi.[23]

Then, extracts of *Stevia rebaudiana* act on a wide spectrum of microorganisms, interestingly their activity is temperature-dependent, in fact increasing the temperature increases also the activity. Taken together, these evidences demonstrate that antimicrobial activity is related to the concentration of secondary metabolites.

8.7.3 Anticancer Activity

An acetone extract of *Stevia rebaudiana* compared with the aqueous extract, showed higher cytotoxic activity on human laryngeal epithelioma cells (HEp2).[38] In an *in vivo* model of Erlisch's Ascites carcinoma in rats, ethanolic extract of *Stevia rebaudiana* showed a good anticancer activity at 300 mg kg^{-1} i.p.,[56] reducing the tumour volume. Stevioside, steviol and also its analogue isosteviol exhibited anticancer activity in the formation of tumour mass, in mouse skin carcinogenesis model as well as inhibited the DNA replication and *in vitro* growth of cancer cells.[50]

8.7.4 Antioxidant and Anti-inflammatory Activity

Antioxidant activity of *Stevia rebaudiana* has been widely documented by several authors in different experimental conditions. Such a profile can justify its involvement in many diseases in which oxidative stress is a predominant factor.[22,37,64] Sutradhar confirmed it in an *in vitro* study in which the common Folin–Ciocalteau method and the aluminium chloride colorimetric method, specific for total flavonoid, have been employed. The antioxidant capacity is correlated with the concentration of flavonoids in the extract.[67]

The most recent evidence came from the study of Tavarini and Angelini.[68] They observed a close correlation among antioxidant potency and the season in which the stevia leaves were collected. In particular, they found that a higher level of phenols was obtained if leaves were collected in spring/summer, and this corresponds to a major antioxidant activity. Indeed, at the basis of antioxidant activity there is the phenolic component, as reported also by Sutradhar and colleagues and in the same year by Singh.[65,67]

In particular, Singh and collaborators carried out an *in vivo* evaluation on mice treated with alloxan, to induce diabetes. They observed that a treatment with *Stevia rebaudiana* extract for 21 d reduced the glucose levels by 39.8% and moreover increased the GSH levels, suggesting that at least in part its beneficial effects may be due to the reduction of the oxidative stress consequent to alloxan-induced damage.[65]

8.8 Toxic and Adverse Effects

Toxicological studies have been widely carried out on stevia and its components, without observing evidence of cancerogenicity or mutagenicity.[52,71,75,79]

In 2015, the FDA completed mutagenotoxicity and genotoxicity studies, concluding that stevioside and rebaudioside A are not genotoxic.[72,73] Several GRAS notices, in particular GRN 253,[24] summarised acute, sub-acute (28 d), sub-chronic (14 weeks) and chronic (22–24 months) treatments with stevioside or rebaudioside A. Indeed, several concentrations were used on rodents in order to assign NOAEL values. In particular, Yamada *et al.* assigned a NOAEL value of 550 mg kg^{-1} d^{-1}, considered as a high dose, calculated in F344 rats feed with 0.1, 0.3 and 1% of SGs for 22–24 months.[79] For stevioside the LD_{50} value is between 8.2–17 g kg^{-1} in mouse, rat and hamster.[48]

Yamada observed no significant effect on spermatogenesis, cell proliferation and tumour formation on rats fed for 22 months with a dose of 1% stevioside.[79]

Some toxicity evidence has been found in pregnant hamsters fed for 6 d with doses of 0.5–1.0 g kg^{-1} d^{-1}. Indeed, high maternal mortality and decrease of live births per litter and mean foetal weight was observed.

Taken together, the GRNs published demonstrate a safety profile of both stevioside and rebaudioside A. They seem to be well-tolerated in normotensive and hypertensive individuals and in diabetic patients following a long-term consumption (GRAS Notices 367, 619, 638).[25–27]

Maki *et al.* demonstrated that a consumption of rebaudioside A at 1000 mg d^{-1} did not produce a clinical change in blood pressure.[47] Nevertheless, two clinical trials on hypertensive patients reported a reduction of blood pressure after long-term treatment with stevioside.[63]

A systematic review of nine studies (including 756 participants) published in 2015 revealed a non-significant difference in systolic blood pressure and of blood lipid panel, whereas a significant reduction of diastolic blood pressure and fasting blood glucose were observed, suggesting that stevioside may be useful for the prevention of cardiovascular risk. However, rebaudioside A does not appear to have any significant effect on cardiovascular risk factors.[53]

The most common adverse reaction of many plants of the *Asteraceae* family is represented by hypersensitivity reaction.[74] The regulatory agencies claim that stevia rarely causes anaphylaxis. In particular individuals with a pre-existing allergy to members of the *Asteraceae* plant family can be more susceptible to anaphylactic symptoms upon exposure to stevia (EFSA 2010).[13] Kimata, in 2007 published a brief communication on two paediatric patients with atopic eczema and known food allergies after either chewing on stevia leaves or ingesting of stevioside powder and water mixture. In both cases, a dexamethasone injection allowed the emergency to be resolved.[33]

Moreover, in 2010, a multidisciplinary group composed of researchers and clinicians concluded that the exposure to stevia or its products is "likely unsafe" for patients with known allergies to members of the *Asteraceae* family.[72]

However, a study of post-marketing pharmacovigilance, from 2008 to 2012, reported adverse clinical effects were rare, among these gastro-intestinal effects comprised 0.5 incidents per 10 million servings, incidents of the SNC (0.33 per 10 million servings) and skin (0.28 per 10 million servings), concluding support for a safety profile of SGs.[34]

A recent examination of phytochemical components of stevia plant revealed that a number of sesquiterpenes is present in stevia extract, and the most abundant include β-caryophyllene (2.1–16.0%), caryophyllene oxide (1.3–18.7%) and spathulenol (13.4–40.9%).[81] Autoxidation products of these volatile compounds have been hypothesised to be responsible for potential allergic reactions. Fortunately, these components are usually present in the essential oil and not in refined SGs products used in the Western market.

References

1. I. Aranda-González, Y. Moguel-Ordóñez, L. Chel-Guerrero, M. Segura-Campos and D. Betancur-Ancona, *J. Med. Food*, 2016, **19**(9), 844.
2. R. Assaei, P. Mokarram, S. Dastghaib, S. Darbandi, M. Darbandi, F. Zal, M. Akmali and G. H. Ranjbar Omrani, *Avicenna J. Med. Biotechnol.*, 2016, **8**(2), 65.
3. E. Brambilla, M. G. Cagetti, A. Ionescu, G. Campus and P. Lingström, *Caries Res.*, 2014, **48**(1), 19.
4. M. Carakostas, I. Prakash, A. D. Kinghorn, C. D. Wu and D. D. Soejarto, Steviol Glycosides, in *Alternative Sweeteners*, ed. L. O'Brien Nabors, Marcel Dekker, New York, NY, USA, 4th edn, 2012, pp. 159–180.
5. M. C. Carakostas, L. L. Curry, A. C. Boileau and D. J. Brusick, *Food Chem. Toxicol.*, 2008, **46**(7), S1.
6. V. Chatsudthipong and C. Muanprasat, *Pharmacol. Ther.*, 2009, **121**, 41.
7. P. Chan, D. Y. Xu, J. C. Liu, Y. J. Chen, B. Tomlinson, W. P. Huang and J. T. Cheng, *Life Sci.*, 1998, **63**(19), 1679.
8. P. Chan, B. Tomlinson, Y. Chen, J. Liu, M. Hsieh and J. Cheng, *Br. J. Clin. Pharmacol.*, 2000, **50**, 215.
9. R. Curi, M. Alvarez, R. B. Bazotte, L. M. Botion, J. L. Godoy and A. Bracht, *Braz. J. Med. Biol. Res.*, 1986, **19**(6), 771.
10. B. De Slavutzky, *J. Verbraucherschutz Lebensmittelsicherh.*, 2010, **5**, 213.
11. J. L. Dorne, K. Walton and A. G. Renwick, *Food Chem. Toxicol.*, 2001, **39**(12), 1153.
12. P. K. Dutta, M. M. T. Razu, M. K. Alam, M. A. Awal and M. Mostofa, *Int. J. Biol. Res.*, 2010, **2**(8), 17.
13. EFSA, *EFSA J.*, 2010, **8**(4), 1537.
14. EFSA, *EFSA J.*, 2015, **13**(12), 4316.
15. COMMISSION REGULATION (EU) 2016/1814 of 13 October 2016 amending the Annex to Regulation (EU) No 231/2012 laying down specifications for food additives listed in Annexes II and III to Regulation (EC) No 1333/2008 of the European Parliament and of the Council as regards specifications for steviol glycoside (E 960), Official Journal of the European Union, 2016.

16. Z. Fan, T. Wen, Y. Chen, L. Huang, W. Lin, C. Yin and W. Tan, *Oxid. Med. Cell. Longevity*, 2016, **2016**, 6362812.
17. G. F. Ferrazzano, T. Cantile and B. Alcidi, *Molecules*, 2015, **21**, E38.
18. F. Gamboa and M. Chaves, *Acta Odontol. Latinoam.*, 2012, **25**(2), 171.
19. C. Gardana, P. Simonetti, E. Canzi, R. Zanchi and P. Pietta, *J. Agric. Food Chem.*, 2003, **51**(22), 6618.
20. J. M. Geuns, P. Augustijns, R. Mols, J. G. Buyse and B. Driessen, *Food Chem. Toxicol.*, 2003, **41**(11), 1599.
21. J. M. Geuns, R. D. Malheiros, V. M. Moraes, E. M. Decuypere, F. Compernolle and J. G. Buyse, *J. Agric. Food Chem.*, 2003, **51**(4), 1095.
22. S. Ghanta, A. Banerjee, A. Poddar and S. Chattopadhyay, *J. Agric. Food Chem.*, 2007, **55**, 10962.
23. S. Gosh, E. Subudhi and S. Nayak, *Int. J. Integr. Biol.*, 2008, **2**, 27.
24. GRAS Rebaudioside A purified from *Stevia rebaudiana* (Bertoni) Bertoni, GRN 253, 2008.
25. GRAS Purified steviol glycosides with rebaudioside A and stevioside as the principal components, GRN 367, 2011.
26. GRAS Exemption Claim for Steviol Glycoside Extract Preparations, GRN 619, 2015.
27. GRAS Notice for High Purity Rebaudioside A, GRN 638, 2016.
28. P. Gupta, N. Gupta, A. P. Pawar, S. S. Birajdar, A. S. Natt and H. P. Singh, *ISRN Dent.*, 2013, **2013**, 519421.
29. M. H. Hsieh, P. Chan, Y. M. Sue, J. C. Liu, T. H. Liang, T. Y. Huang, B. Tomlinson, M. S. Chow, P. F. Kao and Y. J. Chen, *Clin. Ther.*, 2003, **25**(11), 2797.
30. Y. H. Hsu, J. C. Liu, P. F. Kao, C. N. Lee, Y. J. Chen, M. H. Hsieh and P. Chan, *Chin. Med. J.*, 2002, **65**, 1.
31. A. M. Hutapea, C. Toskulkao, D. Buddhasukh, W. Prapin and G. Thirayudh, *J. Clin. Biochem. Nutr.*, 1997, **23**(3), 177.
32. J. A. Khan, S. K. Mishra and K. Vandana, *Int. J. Biol., Pharm. Allied Sci.*, 2012, **1**(10), 1517.
33. H. Kimata, *Allergy*, 2007, **62**, 565.
34. R. L. Kingston, S. Prawer and L. van de Ligt, in *Proceedings of the 7th Stevia Symposium Euprint*, ed. J. M. C. Geuns, Belgium, 2013, pp. 189–200.
35. J. A. Ko, Y. B. Ryu, H. J. Kwon, H. J. Jeong, S. J. Park, C. Y. Kim, Y. J. Wee, D. Kim, W. S. Lee and Y. M. Kim, *Appl. Microbiol. Biotechnol.*, 2013, **97**(18), 8151.
36. E. Koyama, N. Sakai, Y. Ohori, K. Kitazawa, O. Izawa, K. Kakegawa, A. Fujino and M. Ui, *Food Chem. Toxicol.*, 2003, **41**(6), 875.
37. A. Jahan, M. Mostafa, H. Hossain, I. Nimmi, A. Sattar, A. Alim and S. M. Moeiz, *Bangladesh Pharm. J.*, 2010, **13**, 67.
38. S. Jayaraman, M. S. Manoharan and S. Illanchezian, *Trop. J. Pharm. Res.*, 2008, **7**, 1143.
39. JECFA 2010 73rd Meeting – Compendium of food additive specifications. Steviol glycosides, pp. 17–22.
40. JECFA 2016 82nd JECFA –Steviol glycosides. Chemical and Technical Assessment (CTA), pp. 1–10.

41. P. B. Jeppesen, S. Gregersen, C. R. Poulsen and K. Hermansen, *Metabolism*, 2000, **49**(2), 208.
42. V. Ilić, S. Vukmirović, N. Stilinović, I. Čapo, M. Arsenović and B. Milijašević, *Biomed. Pharmacother.*, 2017, **90**, 216.
43. N. Ishima and O. Katayama, *Rep. Natl. Food Res. Inst.*, 1976, **31**, 80.
44. C. N. Lee, K. L. Wong, J. C. Liu, Y. J. Chen, J. T. Cheng and P. Chan, *Planta Med.*, 2001, **67**(9), 796.
45. Y. Liu, T. Sato, B. O'Rourke and E. Marban, *Circulation*, 1998, **97**, 2463.
46. J. Ma, Z. Ma, J. Wang, R. W. Milne, D. Xu, A. K. Davey and A. M. Evans, *Diabetes, Obes. Metab.*, 2007, **9**, 597.
47. K. C. Maki, L. L. Curry, M. C. Carakostas, S. M. Tarka, M. S. Reeves, M. V. Farmer, J. M. McKenney, P. D. Toth, S. L. Schwartz, B. C. Lubin, M. R. Dicklin, A. C. Boileau and J. D. Bisognano, *Food Chem. Toxicol.*, 2008, **46**(7), S40.
48. P. J. Medon, J. M. Pezzuto, J. M. Hovanec-Brown, N. P. Nanayakkara, D. D. Soejarto, S. K. Kamath and A. D. Kinghorn, *Fed. Proc.*, 1982, **41**, 1568.
49. N. Mishra, *J. Nat. Sci. Res.*, 2011, **1**, 3.
50. Y. Mizushina, T. Akihisa, M. Ukiya, Y. Hamasaki, N. C. Murakami, I. Kuriyama, T. Takeuchi, F. Sugarawa and H. Yoshida, *Life Sci.*, 2005, **77**, 2127.
51. A. I. Nikiforov, M. O. Rihner, A. K. Eapen and J. A. Thomas, *Int. J. Toxicol.*, 2013, **32**(4), 261.
52. A. P. Nunes, S. C. Ferreira-Machado, R. M. Nunes, F. J. Dantas, J. C. De Mattos and A. Caldeira-de-Araújo, *Food Chem. Toxicol.*, 2007, **45**(4), 662.
53. I. J. Onakpoya and C. J. Heneghan, *Eur. J. Prev. Cardiol.*, 2015, **22**(12), 1575.
54. J. E. Park and Y. S. Cha, *J. Sci. Food Agric.*, 2010, **90**(7), 1099.
55. D. Preethi, T. M. Sridhar, P. Josthna and C. V. Naidu, *J. Ecobiotechnol.*, 2011, **3**, 5.
56. P. Rajesh, V. Kannan and M. T. Durai, *J. Curr. Biotica*, 2010, **3**, 549.
57. B. Rizzo, L. Zambonin, C. Angeloni, E. Leoncini, F. V. Dalla Sega, C. Prata, D. Fiorentini and S. Hrelia, *Oxid. Med. Cell. Longevity*, 2013, 348169.
58. A. Roberts and A. G. Renwick, *Food Chem. Toxicol.*, 2008, **46**(7), S31.
59. SCF (Scientific Committee on Food), 1984, Reports of the Scientific Committee for Food Concerning Sweeteners (Opinion Expressed by the SCF on 14 September 1984), in Food Science and Techniques. Brussels, Belgium: Commission of the European Communities (EEC), Health & Consumer Protection Directorate-General, Scientific Committee on Food, 1985. (Reports of the Scientific Committee for Food (16th series)).
60. SCF (Scientific Committee on Food), 1989. Reports of the Scientific Committee for Food on Sweeteners. 21th series. Opinion expressed 11 December 1987 and 10 November 1988, and adopted 10 November 1988.
61. SCF (Scientific Committee on Food), 1999. Opinion on Stevioside as a Sweetener (Adopted on17/6/99). Belgium: European Commission, Health & Consumer Protection Directorate-General,Scientific Committee on Food. [CS/ADD/EDUL/167 Final].
62. T. Suanarunsawata, S. Klonganichapaka, S. Rungseesantivanona and N. Chaiyabutrb, *East. J. Med.*, 2004, **9**, 51.

63. N. Sharma, R. Mogra and B. Upadhyay, *Ethno-Med.*, 2009, **3**(2), 137.
64. N. Shivanna, M. Naika, F. Khanum and V. K. Kaul, *J. Diabetes Complications*, 2012, **27**, 103.
65. S. Singh, V. Garg and D. Yadav, *Int. J. Pharmacol. Pharm. Sci.*, 2013, **5**, 297.
66. I. Y. Strobykina, M. G. Belenok, M. N. Semenova, V. V. Semenov, V. M. Babaev, I. Kh. Rizvano, V. F. Mironov and V. E. Kataev, *J. Nat. Prod.*, 2015, **78**, 1300.
67. S. Sutradhar, U. Preeti, A. Humera and M. Muneem, *Res. J. Pharm. Biol. Chem. Sci.*, 2013, **4**, 674.
68. S. Tavarini and L. G. Angelini, *J. Sci. Food Agric.*, 2013, **93**, 2121.
69. L. Testai, I. Strobykina, V. V. Semenov, M. Semenova, E. Da Pozzo, A. Martelli, V. Citi, C. Martini, M. C. Breschi, V. E. Kataev and V. Calderone, *Int. J. Mol. Sci.*, 2017, **18**(10), 1.
70. T. Tomita, N. Sato, T. Arai, H. Shiraishi, M. Sato, N. Takeuchi and Y. Kamio, *Microbiol. Immunol.*, 1997, **41**(12), 1005.
71. K. Toyoda, H. Matsui, T. Shoda, C. Uneyama, K. Takada and M. Takahashi, *Food Chem. Toxicol.*, 1997, **35**(6), 597.
72. C. Ulbricht, R. Isaar, T. Milkin, E. A. Poole, E. Rusie, J. M. Grimes Serrano, W. Weissner, R. C. Windsor and J. Woods, *Cardiovasc. Hematol. Agents Med. Chem.*, 2010, **8**, 113.
73. J. D. Urban, M. C. Carakostas and D. J. Brusick, *Food Chem. Toxicol.*, 2013, **51**, 386.
74. J. D. Urban, M. C. Carakostas and S. L. Taylor, *Food Chem. Toxicol.*, 2015, **75**, 71.
75. L. Xili, B. Chengjiany, X. Eryi, S. Reiming, W. Yuengming, S. Haodong and H. Zhiyian, *Food Chem. Toxicol.*, 1992, **30**(11), 957.
76. D. Xu, S. Zhang, D. J. Foster and J. Wang, *Clin. Exp. Pharmacol. Physiol.*, 2007, **34**, 488.
77. D. Xu, Y. Li, J. Wang, J. Wang, A. K. Davey, S. Zhang and A. M. Evans, *Life Sci.*, 2007, **80**, 269.
78. D. Xu, W. Du, L. Zhao, A. K. Davey and J. Wang, *Planta Med.*, 2008, **74**, 816.
79. A. Yamada, S. Ohgaki, T. Noda and M. Shimizu, *Shokuhin Eiseigaku Zasshi*, 1985, **26**, 169.
80. A. Wheeler, A. C. Boileau, P. C. Winkler, J. C. Compton, I. Prakash, X. Jiang and D. A. Mandarino, *Food Chem. Toxicol.*, 2008, **46**(7), S54.
81. U. Woelwer-Rieck, *J. Agric. Food Chem.*, 2012, **60**, 886.
82. N. L. Zanela, M. F. Bijella and O. P. Rosa, *Pesqui Odontol. Bras.*, 2002, **16**(2), 101.

CHAPTER 9

Steviol Glycosides in Dentistry

M. FRENTZEN*[a], P. BESRUKOW[a,c], A. ACKERMANN[a], S. PIEROG[a], B. SCHIERMEYER[a], J. WINTER[a], U. WÖLWER-RIECK[c] AND D. KRAUS[b]

[a]University of Bonn, Department of Periodontology, Welschnonnenstr. 17, 53111, Bonn, Germany; [b]University of Bonn, Department of Prosthodontics, Welschnonnenstr. 17, 53111, Bonn, Germany; [c]University of Bonn, Institute of Nutritional and Food Sciences, Bioanalytics, Endenicher Allee 11-13, 53115, Bonn, Germany
*E-mail: frentzen@uni-bonn.de

9.1 Introduction

The control of the oral biofilm (plaque) is a fundamental aspect of dental health. Fermentable carbohydrates are implicated in the etiology of plaque associated diseases, such as gingival inflammation and dental decay. Therefore, for a long-time, sugar substitutes were introduced into dental care to replace, *e.g.* sucrose. The cariostatic potential of sweeteners is evidence based.[1,2] Sugar mixtures including sweeteners may also have a potential to affect growth and pH decline in complex oral biofilms.[3] However, based on the fact that sweeteners cause problems in food formations (*e.g.* volume aspects, heat stability) their use is limited to date.

For many years steviol glycosides have been stated to have non-cariogenic effects. This statement is based on the assumption that in the oral cavity steviol glycosides are non-fermentable sugar substitutes. Literature

Food Chemistry, Function and Analysis No. 7
Steviol Glycosides: Cultivation, Processing, Analysis and Applications in Food
Edited by Ursula Wölwer-Rieck

Published by the Royal Society of Chemistry, www.rsc.org

reviews covering the last four decades reveal that there is only little evidence-based knowledge concerning this issue. However, in recent years current studies indicate that stevia extracts might have preventive effects to suppress the growth and activity of oral pathogens linked to caries and gum diseases.[4]

9.2 Oral Diseases

The term "oral diseases" covers pathological disorders in both soft (gingiva) and hard (bone, teeth) tissues of the oral cavity. Oral diseases are not only serious threats to oral health but are also very important to general health and quality of life. Recent research shows the connection between poor oral health and systemic diseases such as diabetes, respiratory diseases, heart diseases and premature birth. Oral diseases range from dental cavities and gum diseases to cancer and dentofacial abnormalities (cleft lip and palate) or trauma from injuries.

Worldwide, 60–90% of school children and nearly 100% of adults have dental cavities. Severe periodontal (gum) disease, which may result in tooth loss, is found in 15–20% of middle-aged (35–44 years) adults. Globally, about 30% of people aged 65–74 have no natural teeth. Risk factors for oral diseases include an unhealthy diet, tobacco use, harmful alcohol use, poor oral hygiene and social determinants.[5]

Oral diseases can be minimised by addressing common risk factors. These include:

- decreasing sugar intake and maintaining a well-balanced nutritional intake to prevent tooth decay and premature tooth loss
- consuming fruit and vegetables that can protect against oral cancer
- stopping tobacco use and decreasing alcohol consumption to reduce the risk of oral cancers, periodontal disease and tooth loss
- ensuring proper oral hygiene
- using protective sports and motor vehicle equipment to reduce the risk of facial injuries
- safe physical environments.

Dental care is a significant economic burden for high-income countries, where 5–10% of public health expenditure relates to oral health. In low- and middle-income countries, public oral health programmes are rare. The high cost of dental treatment may be reduced by effective prevention and health promotion measures.[5] The definite relationship between dietary consumption of sugars and the incidence of dental caries is evidence-based in dentistry. Recent epidemiological studies showed a certain decline in caries prevalence.[6]

For a long period of time sugar substitutes have successfully been used in health care, including dentistry as dietetic sweeteners for many medical reasons.

9.3 Oral Biofilm/Dental Plaque

Dental plaque is an oral biofilm that tightly adheres to the teeth and gums and consists of a diverse species of both bacteria and fungi, embedded in salivary polymers and microbial extracellular products. Oral biofilms can either be in an uncalcified state that can be removed by tooth brushing, or in a calcified state (calculus).

The formation of biofilms on the surfaces of the oral cavity is a precondition to the initiation and progression of all dental hard tissue destruction and gum diseases.[7] Current evidence supports the idea that pathogenic dental plaques do not form spontaneously. Instead, it has long been recognised that plaque formation is a temporally and spatially heterogeneous process, in which cariogenic plaques grow and change in response to environmental pressures.[8] The key events in plaque formation are:[9]

- pellicle formation
- colonisation
- maturation.

Dental caries and periodontal disease should be considered as consequences of ecologically driven imbalances of oral microbial biofilms.[10] These ecologically driven changes in oral biofilms result in increases in the proportions of pathogenic microorganisms, which possess enzymatic and structural determinants that may render them more virulent than organisms associated with oral health. The transition from health to disease is associated with compositional and metabolic changes in populations of bacteria that form the biofilms colonising the hard and soft tissues of the mouth.

Dental caries development is directly correlated with repeated cycles of plaque acidification, which encourage the emergence of aciduric bacteria at the expense of bacteria that are less acidogenic and less acid tolerant.[11] Acid production in dental plaque from the fermentation of dietary carbohydrate leads to, for example, the demineralisation of underlying enamel and dentine (Figure 9.1).

The oral cavity provides a habitat for a wide variety of microbial species. Fluctuations in the conditions in the oral cavity cause rapid changes in availability of nutrient sources and pH values. Despite these apparently unfavourable milieu characteristics, approximately 500 microbial species colonise the oral cavity.[12]

Streptococci from all areas of the oral cavity are isolated and account for a large part of the oral resident flora. In epidemiological studies, streptococci were implied as the main pathogen of:

- enamel caries in children and young adults
- root caries in the elderly population
- and "Early Childhood Caries" in young children.

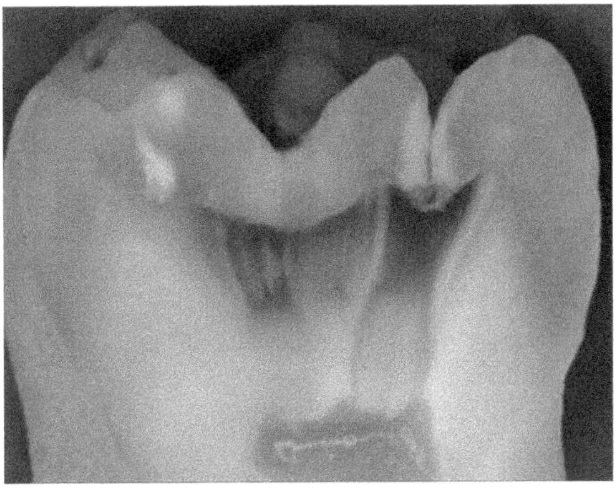

Figure 9.1 Slice of a carious tooth. The white colour indicates demineralisation of the enamel; the underlying dark coloured areas represent dentin caries.

Table 9.1 Different groups of oral streptococci.

Mutans group	Salivarius group	Anginosus group	Mitis group
• *S. mutans* • *S. sobrinus* • *S. rattus*	• *S. salivarius* • *S. vestibularis*	• *S. constellatus* • *S. intermedius* • *S. anginosus*	• *S. sanguis* • *S. gordonii* • *S. parasanguis* • *S. oralis* • *S. mitis*

Oral streptococci can be divided into four types according to Marsh[12] (Table 9.1):

Active carious lesions frequently lack high levels of *S. mutans*, but consistently harbour subpopulations of bacteria, mainly streptococci and lactobacilli, that are highly acidogenic, and more importantly, acid-tolerant.[13] The most relevant characteristic of acid-tolerant species in terms of caries development is that these organisms continue to grow and to produce acids at pH-values well below that required for the dissolution of enamel.[14]

Biofilm formation is favoured by the expression of adhesive glucanes through streptococci species, assembling an exopolysaccharide (EPS) matrix containing bacteria, proteins and enzymes.[15,16] EPS formation is mediated by glycosyltransferases.[17] Acid production is associated with *e.g.* lactobacilli.

S. mutans and *S. sobrinus* are the main pathogens of dental caries and shown to build biofilms.[18–20] Additionally, *S. mutans* has shown to be the most acid-tolerant organism of the dental streptococci. The incidence of *S. mutans* is increased in caries patients. *S. sanguinis* and *S. salivarius* lead to a lesser incidence of *S. mutans* in the oral micro flora, due to secretion of antimicrobial toxins such as hydrogen peroxide and bacteriocins.[21,22] They are components of a caries-free oral flora.[23] Metabolites of *S. salivarius* have been shown to reduce the biofilm formation of *S. mutans.*[24]

9.4 Biofilm Management in Oral Diseases

Oral health care could show that prevention of tooth decay and periodontal diseases may be highly effective, if biofilm management is performed.[25] Different aspects may influence biofilm activity in the oral cavity:

- mechanical and chemical removal of the dental plaque
- application of fluorides
- inhibition of colonisation of tooth surfaces
- antimicrobial drugs
- reduction of fermentable carbohydrates consumption.

Tooth brushing and mouth rinsing, combined with fluoride applications, are the most common and effective tools in preventive dental care. Antimicrobial drugs have only limited impact on dental plaque. Biofilms are more resistant to biocides and medical side effects, as allergic reactions also resistant to pathogens must be taken into consideration. Inhibition of colonisation of oral surfaces in molecular biomedicine is under development. Furthermore, active immunisation against dental plaque or synthetic antibodies to prevent the bacterial colonisation are subjects of basic research.[26,27] Because fermentable carbohydrates are significantly implicated in the etiology of dental caries, sugar substitutes in addition to plaque removal and fluoride applications are the most promising tools in preventative dentistry.

Microbial biofilms on oral surfaces usually exist in dynamic equilibria with host defences and are compatible with maintenance of the integrity of target tissues. The populations consisting of communities of oral microbial residents are not equally distributed throughout the oral cavity. A predominant microflora exists on different sites. Although saliva contains up to 10^8 microorganisms per mL, it is not expected to have a residential microflora. The normal swallowing process prevents the bacteria from reproducing and thus remaining in the oral cavity. The microbial community associated with teeth (biofilm/dental plaque) is composed differently due to the various local environmental conditions on each tooth surface.

Fermentable carbohydrates are implicated in the etiology of biofilm associated oral diseases.[28] Therefore, artificial and natural sweeteners were introduced into preventive dental care to replace sucrose (Tables 9.2 and 9.3). Foods and beverages can be considered safe for teeth, if they are neither cariogenic nor erosive.[29] These "tooth-friendly" properties are, for example, determined in standardised plaque-pH tests.[30]

The main sugar substitutes used are sorbitol and xylitol. Xylitol is considered to be non-cariogenic because it is not fermentable by oral bacteria.

Table 9.2 Examples of non- or low-calories artificial sweeteners. The sweetness intensities are compared to sucrose.

Name	Sweetness intensity
Acesulfame potassium	200
Aspartame	160–200
Salt of aspartame-acesulfame	350
Cyclamate	30
Dulcin	250
Glucin	300
Neotame	8000
Saccharin	300
Sucralose	600

Table 9.3 Examples of non- or low-calories natural sweeteners: the sweetness intensities are compared to those of sucrose.

Name	Sweetness intensity (by weight)
Hydrogenated starch hydrolysates	0.4–0.9
Lactitol	0.4
Isomalt	0.45–0.65
Mannitol	0.5
Glycerol	0.6
Sorbitol	0.6
Erythritol	0.7
Maltitol	0.9
Xylitol	1.0
Mabinlin	100
Stevia	250
Pentadin	500
Curculin	550
Brazzein	800
Thaumatin	2000
Monellin	3000

Since sorbitol in solution can be fermented only slowly by *S. mutans*, chewing sorbitol-sweetened gum does not cause a decrease in plaque pH.[31]

Also, steviol glycosides as natural sweeteners have non-cariogenic effects.[32–34] Moreover, the addition of steviol glycosides to carbohydrate sources may have suppressive effects to the fermentation of sugar.[33]

9.5 Steviol Glycosides in Dental Care

9.5.1 Literature Overview

A detailed database research and literature analysis was performed to evaluate the evidence of steviol glycosides in dental care.

The first scientific publications on "steviosides/rebaudiosides" in the context of "caries" or "dentistry" can be found in medical databases in the late

1970s. They originate from Japan.[35,36] Steviol glycosides have been used as sugar substitutes in Japan for decades. In their *in vitro* microbiological experiments, stevioside, xylitol and sorbitol were tested with *L. plantarum*, *L. casei* and *S. mutans*, but the bacteria could not be cultivated solely with the sugar substitutes. Furthermore, according to the abstract, an inhibition of the bacterial dextransucrase of *S. mutans* and a slightly inhibitory effect on the bacterial invertase of *Clostridium* (species not defined in detail) by stevioside could be determined. However, the methods of these investigations were not published in this abstract.[35]

The effect of sorbitol, xylitol, maltitol, stevioside and thaumatin on *S. mutans in vitro* was studied.[36] In this microbiological assay no acid or glucan formation could be detected for the tested substitutes. The precise experimental setup is not publicly available. Merely the abstracts of the studies were published in English, the whole paper is available exclusively in Japanese.

In 1981, Berry and Henry examined *S. mutans* with sucrose, glucose, fructose and stevioside for 24 h *in vitro*.[37] A lower growth rate and a lower acid production under anaerobic conditions for stevioside was observed when compared to the carbohydrates.

In vitro growth was determined by measurement of optical density and production of acids by monitoring the pH value and total titratable acidity. Growth and acid production of each strain during a 24 h period was determined in triplicate. Growth rates of *S. mutans* was minimal. The acid production in the presence of stevioside was compared to that of sucrose, fructose and glucose statistically reduced. However, the validity of the data is questionable because unfortunately no negative control was performed.

Pinheiro *et al.* could detect an inhibition of the production of extracellular polysaccharides in dental plaque *in vitro* by applying a 0.25% stevia leaf extract.[38] Although stevioside showed a non-significant decrease in plaque fermentation at low concentration (0.25%) the sweetener significantly inhibited the synthesis of insoluble polysaccharides in dental plaque (up to 78%). The entire survey is published in Portuguese, the exact experimental data are not available.

One of the most important studies on this topic is "Evaluation of the Cariogenic Potential of the Intense Natural Sweeteners stevioside and rebaudioside A."[32] It shows *in vivo* that rats, fed with stevioside and rebaudioside A after being infected with *S. sobrinus*, have similarly low bacterial counts for *S. sobrinus* as the negative control.

The oral cavities of sixty rat pups were colonised with *S. sobrinus*. They were divided into four groups that were all fed a basal diet. In group 1 sucrose (1.3%), in group 2 stevioside (0.5%) and in group 3 rebaudioside A (0.5%) were added. Group 4 did not receive any additive (negative control). The pups were sacrificed after five weeks, *S. sobrinus* counts were made, sulcal enamel, dentinal and smooth surface caries was evaluated according to Keyes' technique. The sucrose group (1) had the highest number of *S.*

sobrinus among the four groups. There were no significant differences in *S. sobrinus* counts between the other three groups. Sulcal caries scores were significantly higher in the sucrose-fed rats. No significant differences were found between the remaining groups which had very low incidences of sulcal caries. Dentinal sulcal caries was slightly higher, but not significant in the sucrose-fed animals compared to the two steviol glycoside groups. The animals receiving no additive had no dentin caries. Smooth surface caries was also found only in the group receiving sucrose. The *S. sobrinus* populations in the experimental groups receiving either stevioside or rebaudioside A were comparable to that in group 4, which was not supplied with any additive. This suggests that under the conditions of this study, stevioside and rebaudioside A had no detectable effect on the colonisation and growth of *S. sobrinus* in the oral cavities of the experimental animals compared to the control group. In this study, steviol glycosides act as a pure sugar substitute not modifying the microbial flora. However, the two steviol glycosides were found to be noncariogenic.

The article by Grenby[39] often appears in bibliographies of papers related to the subject "steviosides and caries prevention", but the topic is only included as a review citing a Japanese study.[36]

Zanela *et al.* were unable to demonstrate in their experiment with 200 children that the mouthwash with 0.5% stevioside + 0.05% sodium fluoride causes a significant plaque reduction.[40] In this clinical *in vivo* study the effect of daily mouth rinses on dental plaque accumulation and on salivary *S. mutans* counts was observed over four weeks. The solutions used were mentholated deionised water (group 1), 0.12% chlorhexidine digluconate + 0.05% sodium fluoride (group 2), 0.2% chlorhexidine digluconate (group 3), and 0.5% stevioside + 0.05% sodium fluoride, pH 3, 4 (group 4). The plaque index was measured according to the Löe index at the beginning and at the end of the experiment.[41] The quantification of streptococci was accomplished before the beginning of the mouth rinse, 24 h after the first mouth rinse and a week after the last mouth rinse. Each time 5 mL of the tested solutions was rinsed for 1 min. Group 2 and 3 containing chlorhexidine digluconate presented statistically significant differences, when compared to group 1 and 4. Stevioside solution (group 4) had no inhibitory effect on plaque indices and levels of *S. mutans* in this experiment. The authors assume that the differences were insignificant because of the initially low number of *S. mutans* counts in all participants.

The effects of stevia leaf extract on strains of *S. mutans* grown in Müller-Hinton agar plates and *Lactobacillus acidophilus* were described by Chavarria *et al.* in 2008.[42]

In this *in vitro* study *Stevia rebaudiana* leaves were extracted with water, ethanol, methanol, ethyl acetate or hexanol and afterwards dried by evaporation.

L. acidophilus and *S. mutans* were cultivated for a day on tryptic-soy agar. Tryptic agar (20 mL) were mixed in a test tube with 10 μL of *S. mutans or L. acidophilus* and poured into Petri dishes. Afterwards, 20 μL of different concentrations were placed on the agar, the plates were anaerobically incubated at 37 °C. Growth was observed at 14, 48 and 72 h. Vancomycin and chlorhexidine were used as positive control.

Initially no inhibitory halo could be determined. After 72 h, a certain inhibitory effect on *L. acidophilus* was measured. The results of the publication are extremely scarce and not very detailed. The experiments were carried out in duplicate ($n = 2$). The content of the results should be evaluated critically.

In 2010, Blauth de Slavutzky published a clinical investigation to determine the effect of a 10% stevia leaf extract solution on plaque formation *in vivo*.[43] The number of study participants was relatively low ($n = 8$), the plaque index was measured *in vivo* according to Silness-Löe and O'Leary.[41] Due to the low number of volunteers the statistical evaluation has to be interpreted critically. The accumulation of dental plaque after rinsing with stevia was up to 82% less than with sucrose measured by Silness-Löe index and up to 40% less when measured by O'Leary index. According to the material and methods, no negative, only one positive control with sucrose was carried out. The authors show in their experiment a lower incidence of plaque formation with stevia leaf extract compared to sucrose.

In the double blind study of Rumiantsev *et al.*,[44] chewing gum with xylitol, sorbitol and with 1.25% and 2.5% stevioside, as well as placebo gums, were tested by 126 medical students. The full text has only been published in Russian, therefore some questions about the methodology of this *in vivo* study remain open. Stevioside-containing chewing gums behaved similarly to chewing gums with xylitol and sorbitol.

The effects of stevia leaf extract on 16 different streptococci and lactobacilli species have been investigated by Gamboa and Chaves.[45] They dissolved the stevia extract in hexane, methanol, ethanol, ethyl acetate and chloroform. The experimental set up is similar to that of Chavarria *et al.* from 2008.[42] There were negative controls with vancomycin and azithromycin. An inhibitory effect could be observed in all samples, which is not surprising, since all the substrates tested have an antibacterial effect caused by the solvent. How a special effect caused by the leaf extract should be determined is questionable. Thus, this study is also not fully indicative for antibacterial effect of stevia leaf extract.

The effect of stevia on the formation of intra and extracellular polysaccharides, vital cell numbers, cell mass and reduction of enamel hardness under *S. mutans* biofilms in the *in vitro* experiments of Giacaman *et al.* is similar to other sugar substitutes.[46] *S. mutans* biofilms were cultured on bovine enamel slabs and exposed to 0.9% NaCl (caries-negative control), stevia, sucralose, saccharin, aspartame or fructose and 10% sucrose (caries-positive control) three times a day for 5 min. After 5 d the biofilms were recovered. All tested sweeteners (except fructose) showed less enamel demineralisation than sucrose, but still higher than the negative control. Stevia, sucralose and saccharine reduced the number of vital cells when compared with sucrose. They also reduced the formation of extracellular polysaccharides in comparison to the positive control at levels similar to the negative control.

Brambilla *et al.* showed in 2014 that 10% solutions of stevioside and rebaudioside A behave differently than the positive control with sucrose.[34] The effect of stevia extracts on *in vitro* biofilm formation and *in vivo* plaque pH was evaluated. Plaque pH and the biofilm formation of *S. mutans* are significantly lower than in the positive control. There were no negative controls. A wild strain of *S. mutans* isolated from human dental plaque samples was used for the *in vitro* investigation on biofilm formation. Three solutions containing 10% stevioside, 10% rebaudioside A and 10% sucrose were tested. The microbial count *in vitro* was measured by MTT assay. For the *in vivo* testing, 20 volunteers rinsed with each solution for 1 min. The plaque pH was measured after 5, 10, 15, 30, 45 and 60 min. The subjects were asked to refrain from oral hygiene procedures for 24 h and from using tobacco, eating and drinking anything except water for 2 h before the test. After 5, 10, 15 and 30 min, the sucrose rinse produced a statistically significantly lower pH value compared to the rinses containing stevioside and rebaudioside A. After 45 min, the pH showed no significant difference in levels among the three solutions.

Giongo *et al.* surveyed sugar substitutes containing a combination of lactose and stevioside *in vivo* in an experiment with nine participants.[47] They refrained from brushing their teeth for three days. On the fourth day pH was measured intraoral by the strip pH-method. Interproximal plaque pH was measured before and up to 60 min after a 10 mL mouth rinse for 1 min with the test solutions. The mouth rinses of the groups contained 93% lactose + 7% stevioside (1); 6.8% saccharin + 13.6% cyclamate + 0.82% stevioside (2); 18% sucrose (3); mineral water as negative control (4) and 93% lactose solution (5). Statistically, there were no significant differences to the positive control with sucrose. The critical pH <6.5 for loss of minerals in dentine was reached in all samples. In the presented combination, the formulations containing sweeteners (93% lactose + 7% stevioside and 6.8% saccharin + 13.6% cyclamate + 0.82% stevioside) were cariogenic.

Ajagannanavar *et al.* showed that stevia leaf extract dissolved in 70% ethyl alcohol has an inhibitory effect on *S. mutans* and *L. acidophilus*, which is lower than the inhibition by chlorhexidine in the positive control and greater than that of the negative control group with water.[48] The authors note critically that the antibacterial effect could also be caused by the use of the ethanol.

In addition to these microbiological tests, animal experiments and clinical studies reviews have been published. Contreras (2013) and Ferrazzano *et al.* (2015) surveyed the published literature on stevioside.[49,50] They discussed the data concerning the anti-bacterial role of steviol glycosides on the oral bacterial flora. The data shown in their papers provide evidence that extracts from *S. rebaudiana* are not cariogenic. The authors suggest future research should be focused on *in vivo* studies to evaluate the effects on the development of dental caries. It basically agrees with the results described in this literature review. Although there is a lot of literature on steviol glycosides, there is little systematic research with impact in preventive dentistry. In the literature given the influence, on the oral microbiome and biofilms information is often not described according to today's standards. Some investigations raise questions.

In conclusion: the cariostatic effects of steviol glycosides were only documented in two international publications (Berry *et al.*, 1981 and Das *et al.*, 1992) being performed according to recent scientific standards. These publications indicate a reduction in growth and decrease of acid production by *S. mutans* (*in vitro*) and reduction of caries in rats infected with *S. sobrinus*.

9.5.2 Current Research

The existing literature concerning the impact of stevia extracts on dental health implies an antibacterial effect of steviol glycosides, with some results being inconclusive, though. Therefore, current investigations concentrate on the effects of stevia extracts predominantly steviol glycosides—on preventive dentistry, considering a series of parameters that are related to the etiology of oral diseases, *e.g.* tooth decay.

In this context, three commercially available stevia substrates, namely "RebA" (rebaudioside A content of ≥ 97%), "Stev" (a steviol glycosides mixture, purity ≥ 95%, with a rebaudioside A content of minimum 60%) and "Instant" (a water extraction of stevia leaves containing approximately 28% steviol glycosides), were used in comparison to sucrose in various experiments in order to (a) monitor bacterial growth rates, (b) evaluate sugar consumption and thus acid production, (c) assess the influence on biofilm formation and (d) investigate a potential decomposition of steviol glycosides by the oral microbiome. All experiments based on the *in vitro* cultivation of four oral streptococci strains (*S. mutans, S. sobrinus, S. sanguis, S. salivarius*—all isolated from human caries) in mono and mixed cultures. These streptococci were chosen, because they represent both main pathogen and commensal bacteria and, thus, simulate the actual microbial composition in the oral cavity most likely. In order to breed the microorganisms, brain-heart-infusion (BHI) medium was used, since it ensures optimal conditions for bacterial growth.[51] Furthermore, BHI medium was used to solve and dilute the above-named stevia extracts and sucrose. To simulate realistic conditions for the application of stevia in beverages and food, the sweetness of the given substrates was evaluated in a pre-test. All six testing persons claimed, "Reb A" and "Stev" having a pleasant sweetness at a concentration of 0.5%. Both substances are purified extracts, consisting purely of steviol glycosides. The substrate "Instant"—as a natural, unpurified water extraction of dried stevia leaves—consists up to 28% steviol glycosides plus other plant ingredients such as chlorophyll, polyphenols *etc.*

Regardless of a minor sweetness, the substrate "Instant" was used likewise to all other substrates at a concentration of 0.5% in the following experiments to ensure a clear comparison.

9.5.3 Modulation of Bacterial Growth

After breeding in BHI, the four bacteria mono cultures and one mixed culture of these strains were cultivated in five different media, namely "BHI" (pure BHI medium) as a reference, "Suc" (BHI + 0.5% Sucrose), "RebA" (BHI + 0.5% Rebaudioside A), "Stev" (BHI + 0.5% steviol glycoside) and "Inst" (BHI + 0.5% Instant) for 24 h. Within this period, planktonic growth was monitored measuring the optical density (OD) with a photometer at 600 nm hourly for the first 8 h and once after 24 h. The optical density is a parameter for the number of bacteria cells in a liquid medium. The more turbid the medium is, the higher is the optical density and, consequently, the number of bacteria cells in the medium. In order to exclude the possibility of undesired contamination, a "Blank" control, which was not inoculated with bacteria but contained pure BHI, was run simultaneously and treated equally to the other five media. Figure 9.2 shows the optical density progress of different media with the streptococci mixed culture within 24 h.

Streptococci cultivated in BHI medium with the addition of sucrose ("Suc") showed the fastest planktonic growth, which is affiliated to the energy provided by the carbohydrate sucrose that ensures a prosperous growth. However, even in BHI medium without any addition ("BHI"), bacterial growth was evident. This is due to the medium's compounds: Besides bovine brain- and

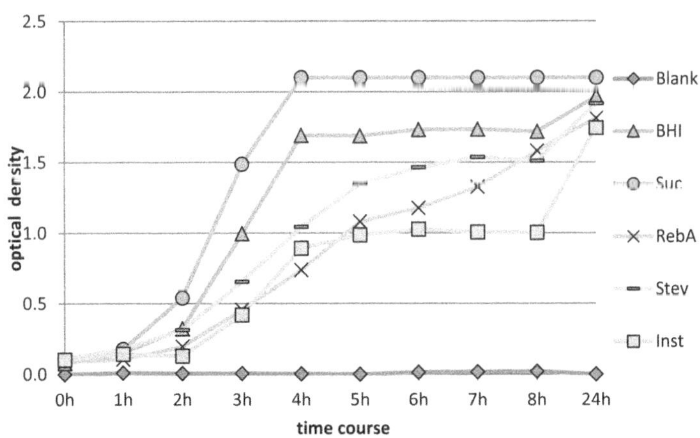

Figure 9.2 Planktonic growth of a streptococci mixed culture in "BHI" (pure BHI medium) as a reference, "Suc" (BHI + 0.5% sucrose), "Reb A" (BHI + 0.5% rebaudioside A), "Stev" (BHI + 0.5% steviol glycoside) and "Inst" (BHI + 0.5% Instant). The addition of sucrose to BHI media accelerated bacterial growth; steviol glycoside addition—especially "Inst" which contains other phytochemicals delays bacterial growth.

heart-infusion, BHI medium contains 0.2% glucose that—like sucrose—can be metabolised and generate ATP (energy) by bacteria.

Both "Suc" and "BHI" show a saturation of growth after 4 h, being on different OD values levels, though. While "BHI" growth stops at OD 1.6, extra sucrose enables a faster and higher growth of "Suc" to OD 2.1, presumably even higher. Due to the experimental setup, OD > 2.0 were not shown on the photometer because of its cut-off at an OD = 2.0. Therefore, all values beyond 2.0 were not taken into consideration and were recorded as OD 2.1.

Bacteria in all stevia substrate media had a significantly decreased/lower growth compared to "Suc" and "BHI". "RebA" and "Stev" show a similar growth rate of the Streptococci being approximately at half the level of "Suc". "Inst" medium, however, had the strongest inhibition of growth despite a lower steviol glycoside concentration. This effect could be affiliated to the existence of further phytochemicals *e.g.* polyphenols, which might have an anti-bacterial effect.

All in all, sucrose accelerated bacterial growth, stevia substrates delayed it. This effect was observed in both mixed and mono cultures. Exemplary for mono cultures, Figure 9.3 shows the optical density progress of *S. mutans*.

9.5.4 Effects of Steviol Glycosides on Bacterial Sugar Consumption and Acid Production

Within bacterial glycolysis, carbohydrates such as sucrose and glucose are metabolised under anaerobic conditions into lactic acid, in parallel converting ADP into ATP, which can be used as an energy source for growth.

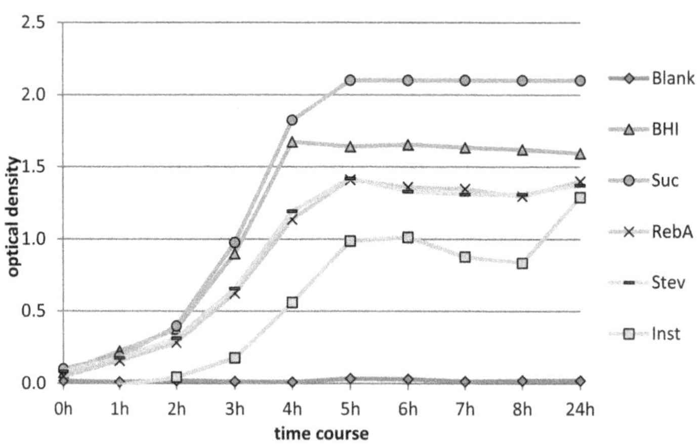

Figure 9.3 Planktonic growth of *S. mutans* in "BHI" (pure BHI medium) as a reference, "Suc" (BHI + 0.5% sucrose), "Reb A" (BHI + 0.5% rebaudioside A), "Stev" (BHI + 0.5% steviol glycoside) and "Inst" (BHI + 0.5% Instant). A delay of bacterial growth is observed when using steviol glycoside formulations instead of sucrose in BHI medium, while "Reb A" and "Stev" virtually having the same effect.

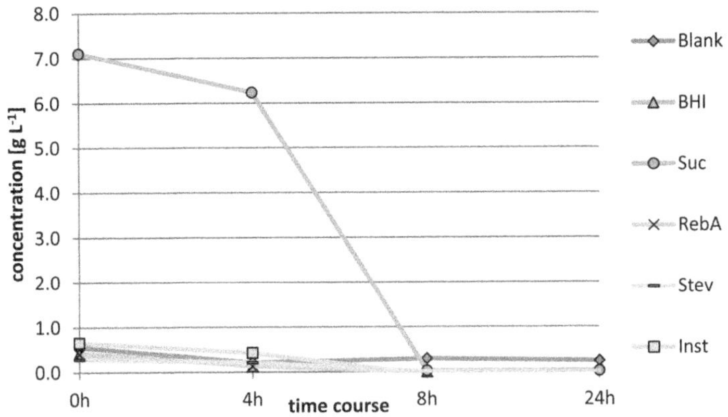

Figure 9.4 Sucrose concentration during 24 h of bacterial growth (mixed culture) in "BHI" (pure BHI medium) as a reference, "Suc" (BHI + 0.5% sucrose), "Reb A" (BHI + 0.5% rebaudioside A), "Stev" (BHI + 0.5% steviol glycoside) and "Inst" (BHI + 0.5% Instant). Sucrose is metabolised in "Suc" (the only medium containing sucrose) steadily. After 8 h no more sucrose is left in "Suc", while no change in sucrose concentration in other media is observed.

Therefore, besides monitoring the optical density, bacterial growth was verified by analysing the sugar and lactic acid concentrations. The latter was also used as an indicator for the pH values measured in the media parallel to the OD.

First, aliquots of media were taken right after inoculation and after 4, 8 and 24 h of bacteria cultivation. These aliquots were used for the following investigations concerning their amount of sucrose, glucose and lactic acid. All parameters were analysed enzymatically.[52] Figures 9.4 and 9.5 show the progress of sucrose (Figure 9.4) and glucose concentration (Figure 9.5) within the tested media.

Since only "Suc" contained sucrose, all other graphs in Figure 9.4 start at a concentration of approx. 0.02% (which represents the standard deviation in this experiment) and remained at this level for 24 h. The sucrose concentration in the medium "Suc" stays at a high level for the first 4 h, then decreases to 0% after 8 h.

With regards to the glucose concentrations in Figure 9.5, it becomes evident that all media start at approximately 2 g glucose per L that comes from the BHI. While glucose concentrations in "RebA", "Stev" and "BHI" decrease fast being at 0% after 8 h, the consumption of glucose in "Inst" is delayed. Hence, after 24 h there is still some glucose in the medium. This delay might be explained by the presence of phytochemicals like polyphenols, which are not present in "RebA" and "Stev". In this case, glucose consumption is evidently restricted by stevia substrates, however only by "Inst" and not by pure steviol glycosides. Glucose consumption in "Suc" is delayed until 8 h. Given the results in Figure 9.4 it becomes obvious that microorganisms use

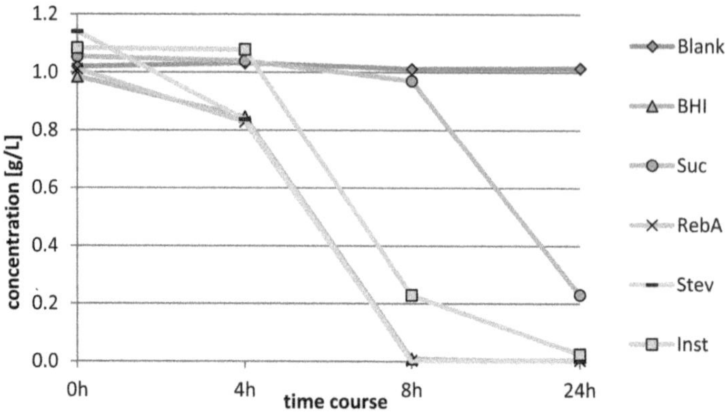

Figure 9.5 Glucose concentration during 24 h of bacterial growth (mixed culture) in "BHI" (pure BHI medium) as a reference, "Suc" (BHI + 0.5% sucrose), "Reb A" (BHI + 0.5% rebaudioside A), "Stev" (BHI + 0.5% steviol glycoside) and "Inst" (BHI + 0.5% Instant). Steady decrease in glucose concentration in "BHI", "Reb A", "Stev" and "Inst". Decomposition of glucose in "Suc" starts after 8 h due to the available sucrose in this medium.

the available sucrose as an energy source until it is depleted in the medium. From this point on, glucose is used by the bacteria to ensure growth. This shows a bacterial preference for sucrose over glucose.

Carbohydrate consumption, either sucrose or glucose, results in the production of lactic acid by the bacteria. As displayed in Figure 9.6, acid production correlates with the consumption of sugars. The more sugars are consumed, the more lactic acid is produced. In this context, media containing stevia substrates are similar to the reference "BHI". Likewise, to OD (Figure 9.2), values of lactic acid are about twice as high in media containing sucrose. However, compared to the reference, there is no obvious influence of stevia substrates on lactic acid production.

In the oral cavity, lactic acid is the main parameter that produces a pH-decline, which is connected to the demineralisation of dentine and, thus, to caries. Therefore, the development of pH was measured in the aforementioned media using a pH-meter. Figure 9.7 shows the results of the pH values during cultivation in "RebA", "Stev", "Inst", "Suc" and "BHI" as reference.

At the start, all pH values in the media were approximately 7.3 with the exception of "Inst", which due to natural leaf components is slightly lower. The sucrose medium has the fastest pH-decline, while the pH value in the media with stevia substrates decreases but is delayed by hours, even compared to the reference.

Investigations concerning planktonic growth of oral streptococci, linked to their sugar consumption, lactic acid production and pH-decline in the respective media show an influence of the application of stevia substrates in substitution to sugars on bacterial metabolism. First, it became obvious that

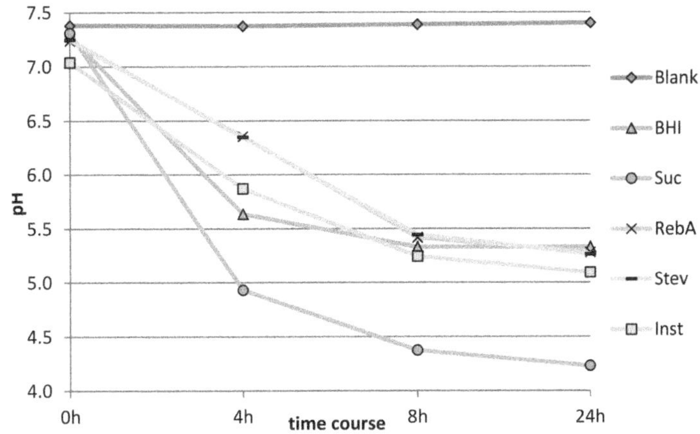

Figure 9.6 pH progress of different media ("BHI" (pure BHI medium) as a reference, "Suc" (BHI + 0.5% sucrose), "Reb A" (BHI + 0.5% rebaudioside A), "Stev" (BHI + 0.5% steviol glycoside) and "Inst" (BHI + 0.5% Instant)) during cultivation of streptococci mixed culture. Decrease of pH is expedited when sucrose is added ("Suc"), while addition of steviol glycoside formulation retards a pH drop during the first 8 h slightly, being at the same level with the reference ("BHI") after 24 h, but still pH 1.5 higher than sucrose.

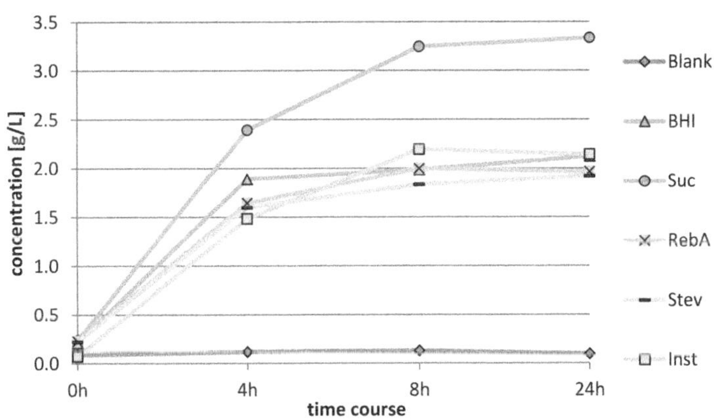

Figure 9.7 Concentration of L-lactic acid during 24 h cultivation of streptococci mixed culture in "BHI" (pure BHI medium) as a reference, "Suc" (BHI + 0.5% sucrose), "Reb A" (BHI + 0.5% rebaudioside A), "Stev" (BHI + 0.5% steviol glycoside) and "Inst" (BHI + 0.5% Instant). Lactic acid production increases and fastens when sucrose is added to BHI medium; adding steviol glycoside formulations to BHI, lactic acid production is slightly inhibited within the first 4 h being at the same level after 24 h.

sucrose accelerates planktonic growth, lactic acid production and pH-decline, all of which favor the formation of oral diseases. Second, in the stevia substrates, bacterial growth and pH-decline were significantly restricted and delayed compared to the reference. Thus, such restriction could not be evident in the experiments concerning sugar consumption and lactic acid production. Obviously, there is an influence on bacterial growth, which does not affect bacterial glycolysis directly. Here, further investigations on bacterial metabolism are needed. The two samples containing more than 95% steviol glycosides—mainly Rebaudioside A—showed very similar results. "Instant" partly showed more restrictive effects on bacterial growth. As a water extract from stevia leaves—"Instant" only contains about 28% of steviol glycosides and 72% of further constituents from the stevia leaves that might have similar or stronger effects. In this case, further studies concerning various phytochemicals in *S. rebaudiana* would be appropriate in order to evaluate the meaning of these results.

9.5.5 Suppression of Oral Biofilm Formation

With *S. mutans* being the main pathogen responsible for plaque in oral cavity, its biofilm formation was investigated with a crystal violet staining protocol in microplates. In a first step, planktonic growth was measured at an optical density of 600 nm. Afterwards, biofilm formation at the wall of the microplate wells was stained with crystal violet. After excessive washing, the incorporated dye was eluted by ethanol and biofilm formation was quantified by measuring the optical density of the eluate at 630 nm. Additionally, to estimate biofilm formation in regard to planktonic growth, the ratio between biofilm accumulation and planktonic growth per well was calculated by dividing the optical densities at 630 nm through the corresponding values at 600 nm wavelength. The experimental setup is visualised in Figure 9.8.

Again, "RebA", "Stev" and "Inst" were used in comparison to sucrose. However, different concentrations of the stevia substrates were used in order to investigate a concentration dependency on biofilm formation. To verify the effect of stevia extracts on biofilm formation in the presence of glucose (from BHI) and sucrose, all media was added 0.5% sucrose solution. Figure 9.9 shows the results of the assay.

Using pure BHI medium biofilm formation hardly occurred. Similar results were obtained adding different formulations of steviol glycosides. When sucrose was added to BHI, the highest biofilm formation was observed. The addition of stevia extracts to medium containing 0.5% sucrose reduced biofilm formation. The addition of "Reb A" inhibited biofilm formation for approximately 20% regardless of the concentration used. The same effect was observed when adding 0.1% of "Stev" and "Instant" to the 0.5% sucrose medium. The addition of 0.5% or 1% of "Stev" and "Instant" to the 0.5% sucrose medium decreased the biofilm formation significantly in arrange of down to 5%.

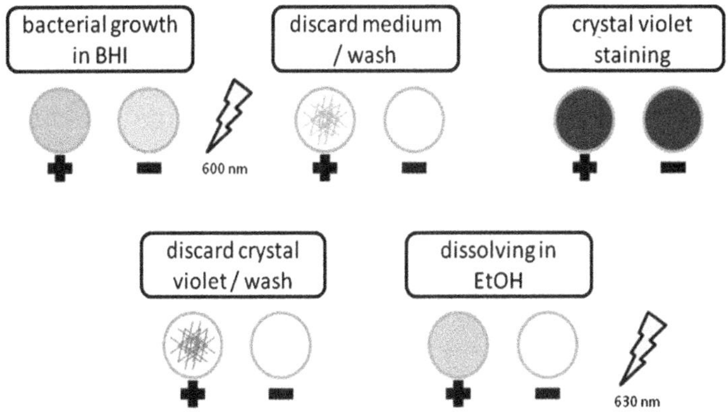

Figure 9.8 Schematic representation of the biofilm *in vitro* assay and its single steps including measurements of optical density, showing the positive (+) and a negative control (−) (Pierog *et al.*, 2015).[53]

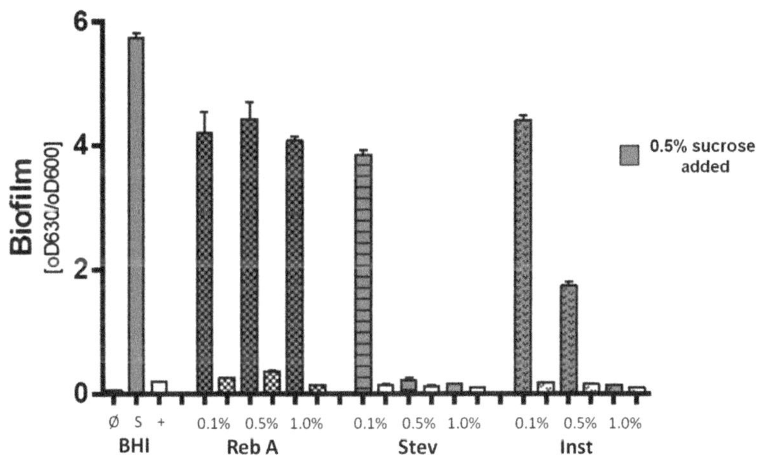

Figure 9.9 Biofilm formation measured in a microplate assay by calculating the ratio of planktonic growth and dyed biofilm at 600 nm and 630 nm ($n = 12$), respectively, in "BHI S" (BHI + 0.5% sucrose) as a reference, "BHI +" (pure BHI), "Reb A" (BHI + rebaudioside A in concentrations of 0.1%, 0.5% and 1.0%), "Stev" (BHI + steviol glycoside in concentrations of 0.1%, 0.5% and 1.0%) and "Inst" (BHI + Instant in concentrations of 0.1%, 0.5% and 1.0%) (Ø = negative control). High biofilm formations are observed when sucrose is added to BHI media. Addition of steviol glycoside formulations inhibited biofilm formation both when added to sucrose and when substituted sucrose. All columns have a significant ($p < 0.05$) restriction of biofilm formation (one-sided ANOVA-test) compared to the reference "BHI S".

This data indicates a suppression of biofilm formation by steviol glycosides, especially stevioside, which was available in "Stev" and "Inst". Furthermore, likewise to the results in experiments concerning planktonic growth, Instant medium showed similar results to "Stev" despite a lower concentration of steviol glycosides, which, again, might be affiliated to other phytochemicals available in the leaves of *S. rebaudiana*.

9.5.6 Decomposition of Steviol Glycosides by Oral Bacteria

Steviol glycosides are the subject of many investigations concerning their sweetness, therapeutic benefits and pharmacokinetics. The latter mainly focus on their degradation by intestinal microorganisms, while steviol glycosides as natural sweeteners are initially exposed to the oral microflora during ingestion. Therefore, the aim of this experiment was to investigate whether oral streptococci are capable to decompose/metabolise steviol glycosides.

After being grown in a brain-heart-infusion medium under optimal conditions, oral streptococci in mono and mixed culture were exposed to three different stevia substrates: "Reb A", "Stev" and "Instant", each of them at three concentrations (0.1%, 0.5%, 1.0%) for 24 h. After 0, 4, 8 and 24 h, the concentration of Reb A was analysed *via* RP-HPLC-UV. The parameters used within the analysis are listed in Table 9.4.

Due to experimental setup, aliquots at 0 h were taken approximately 5 min after inoculation with bacteria. During this short period, some steviol glycosides were already decomposed. Therefore, there were a difference between the concentration analysed in media before inoculation and the concentration at "0 h". Since the substrates differed in their amount of Reb A, results are shown in percentage, in order to visualise the decomposition in a comparative way.

Figure 9.10 shows the results of the analysis of Reb A decomposition.

In general, the absolute quantity of Reb A that was metabolised by oral streptococci in the tested media within 24 h was marginal, but evident. The effects of degradation of steviol glycosides in the oral cavity are unknown so far. Possible metabolites in the oral cavity are unidentified.

Table 9.4 Parameters used for the analysis of steviol glycosides concentration (modified according to Oehme *et al.*, 2013).

Column	Purospher® STAR RP-18 endcapped (3 µm) Hibar® RT 250 × 4, 6 mm
Mobile phase	85% acetonitrile/15% water
Analysis time	15 min
Flow rate	1 mL min^{-1}
Oven temperature	50 °C
Detection	Prostar 325 DAD
Injection volume	10 µL

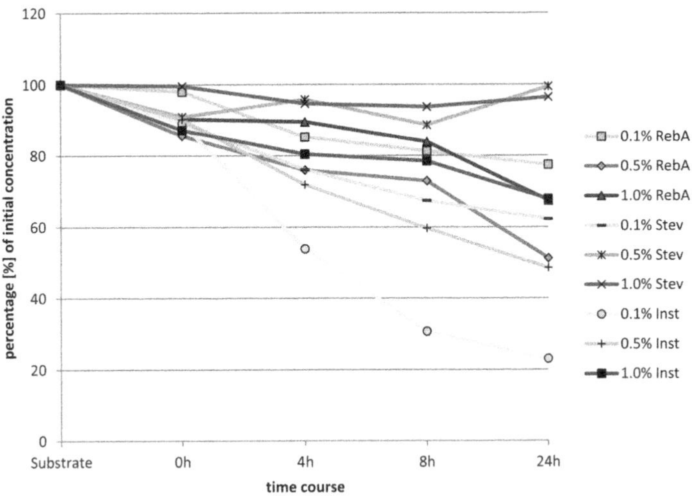

Figure 9.10 Decomposition of rebaudioside A by streptococci mixed culture in "Reb A", "Stev" and "Inst" regarding different concentrations in media, presented in % of initial concentration (=100%). A marginal decrease of rebaudioside A concentration is observed in all media, being most evident in media with low steviol glycoside concentrations (0.1%).

A concentration-dependent degradation could be observed in all tested formulations. The higher the concentration of steviol glycosides in the medium, the relatively less steviol glycosides were decomposed by the streptococci. An explanation might be the limited number of bacterial cells. Further investigations on steviol and the various steviol diterpenes are suggested in order to assess the possible toxicity associated to oral decomposition.

Regarding these investigations, steviol glycosides featured different inhibitory effects on bacterial growth, sugar decomposition and acid production. Decomposition of Reb A and stevioside by oral microorganisms was marginal. Moreover, an inhibitory influence on biofilm formation was evident, even at partial substitution of sugars through steviol glycosides. Consequently, the potential of sugar mixtures including steviol glycosides in food and beverage formulations may be a highly significant approach in preventive dental care. Further scientific verification is suggested in order to understand the impact of steviol glycosides on bacterial metabolism and, thus, on oral health more profoundly.[53,54]

9.6 Interaction of Steviol Glycosides to Oral Tissues

A detailed database research and literature analysis resulted in a lack of references regarding the effects and interactions of steviol glycosides in oral tissues. However, our own activities investigate steviol glycosides-induced impacts on oral tissues. One of the main research topics focuses on oral tumour biology. Interestingly, there are some reports in literature that have been shown that solvent extracts of *S. rebaudiana* have probably anti-tumour potential.[55–60] It could be demonstrated that in preliminary pilot experiments, stevia extracts negatively affected the proliferation rates of human oral squamous cell carcinoma cells. These first obtained data might serve as a promising start for future research attempts on this very thrilling project.

9.7 Perspectives

Beside dietary aspects (zero-calorie sweeteners) under *in vitro* conditions suppressive effects in oral biofilms of stevia extracts have to be discussed. This would represent a positive approach to *e.g.* using antiseptic mouth washes and appears as more natural, targeted and gentle to maintain a positive microflora balance. Even in mixtures with carbohydrates those effects appear in current studies. Such a benefit of sugar mixtures including stevia extracts in food and beverage formulations may be a highly significant approach in preventive dental care. The current *in vitro* evidence emphatically encouraged to perform clinical studies to evaluate the clinical impact of stevia formulations.

References

1. K. A. Ly, P. Milgrom and M. Rothen, *Pediatr. Dent.*, 2006, **28**(2), 154.
2. C. Van Loveren, *Caries Res.*, 2004, **38**(3), 286.
3. E. Söderling, J. Talonpoika and K. K. Mäkinen, *J. Dent. Res.*, 1987, **95**(5), 405.
4. M. Frentzen, S. Bröring and D. Kraus, *The World of Food Ingredients*, 2015, 34.
5. World Health Organization, Oral health, 2012, Fact sheet N°318.
6. W. Micheelis and U. Schiffner, Fourth German Oral Health Study (DMS IV) D.Ä.V., 2006.
7. L. Sbordone and C. Bortolaia, *Clin. Oral Investig.*, 2003, **7**(18), 1.
8. E. Theilade, *Ann. R. Australas. Coll. Dent. Surg.*, 1989, **10**, 62.
9. A. A. Scheie and F. C. Petersen, *Crit. Rev. Oral Biol. Med.*, 2004, **15**, 4.
10. D. J. Bradshaw, A. S. McKee and P. D. Marsh, *J. Dent. Res.*, 1989, **68**, 1298.
11. R. G. Quivey Jr., W. L. Kuhnert and K. Hahn, *Adv. Microb. Physiol.*, 2000, **42**, 239.
12. P. Marsh and M. V. Martin, *Oral Microbiology*, Reed Educational and Professional Publishing Ltd., 4th edn, 1999.
13. J. Van Houte, *J. Dent. Res.*, 1994, **73**, 672.
14. G. R. Bender, S. V. Sutton and R. E. Marquis, *Infect Immun.*, 1986, **53**(33), 1.
15. A. P. Erdem, E. Sepet, G. Kulekci, S. C. Trosola and Y. Guven, *Int. J. Med. Sci.*, 2012, **9**(2), 129.
16. M. I. Klein, J. Xiao, B. Lu, C. M. Delahunty, J. R. Yates and H. Koo, *PLoS One*, 2012, 7(9), 45795.

17. W. H. Bowen and H. Koo, *Caries Res.*, 2011, **45**(1), 69.
18. W. J. Loesche, *Microbiol. Rev.*, 1986, **50**(4), 353.
19. K. G. Babaahmady, S. J. Challacombe, P. D. Marsh and H. N. Newman, *Caries Res.*, 1998, **32**(1), 51.
20. M. Dinis, D. Tavares, I. Veiga-Malta, A. J. M. M. Fonseca, E. B. Andrade, G. Trigo, A. Ribeiro, A. Videira, A. M. S. Cabrita and P. Ferreira, *J. Infect. Dis.*, 2009, **199**(1), 116.
21. C. S. Ryan and I. Kleinberg, *Arch. Oral Biol.*, 1995, **40**(8), 753.
22. J. Kreth, Y. Zhang and M. C. Herzberg, *J. Bacteriol.*, 2008, **190**(13), 4632.
23. P. W. Caufield and A. L. Griffen, *Pediatr. Clin. North Am.*, 2000, **47**(5), 1001.
24. A. Ogawa, S. Furukawa, S. Fujita, J. Mitobe, T. Kawarai, N. Narisawa, T. Sekizuka, M. Kuroda, K. Ochiai, H. Ogihara, S. Kosono, S. Yoneda, H. Watanabe, Y. Morinaga, H. Uematsu and H. Senpuku, *Appl. Environ Microbiol.*, 2011, **77**(5), 1572.
25. I. L. Chapple, P. Bouchard, M. G. Cagetti, G. Campus, M. C. Carra, F. Cocco, L. Nibali, P. Hujoel, M. L. Laine, P. Lingstrom, D. J. Manton, E. Montero, N. Pitts, F. Rangé, N. Schlueter, W. Teughels, S. Twetman, C. Van Loveren, F. Van der Weijden, A. R. Vieira and A. G. Schulte, *J. Clin. Periodontol.*, 2017, **44**, 39.
26. G. Hajishengallis and S. M. Michalek, *Oral Microbiol. Immunol.*, 1999, **14**, 1.
27. M. B. Kupper, M. Huhn, H. Spiegel, J. K. C. Ma, S. Barth, R. Fuscher and R. Finnern, *BMC Biotechnol.*, 2005, **5**, 4.
28. P. Moynihan and P. E. Petersen, *Public Health Nutr.*, 2004, **7**(1A), 201.
29. B. Guggenheim, *Adv. Dent. Res.*, 1995, **9**(2), 129.
30. A. R. Firestone, T. Imfeld, S. Schiffer and F. Lutz, *Caries Res.*, 1987, **21**(6), 555.
31. W. M. Edgar, *Br. Dent. J.*, 1998, **184**, 29.
32. S. Das, A. K. Das, R. A. Murphy, I. C. Punwani, M. P. Nasution and A. D. Kinghorn, *Caries Res.*, 1992, **26**, 363.
33. A. Ackermann and M. Frentzen, *Proceedings of the 5th Stevia Symposium*, ed. J.M.C Geuns, Euprint, Heverlee, Belgium, 2011, p. 145.
34. E. Brambilla, M. G. Cagetti, A. Ionescu, G. Campus and P. Lingström, *Caries Res.*, 2014, **48**, 19.
35. M. Yabu, M. Takase, K. Toda, K. Tanimoto and A. Yasutake, *Hiroshima Daigaku Shigaku Zusshi*, 1977, **9**(1), 12.
36. T. Ikeda, H. Okada and R. Motada, *Nihon Univ. J. Oral Sci.*, 1978, (4), 24.
37. C. W. Berry and C. A. Henry, [Poster ORCA 1981], *J. Dent. Res.*, 1981, **60**, 430.
38. C. E. Pinheiro, S. S. de Oliveira, S. M. da Silva, M. I. Poletto and C. F. Pinheiro, *Rev. Odontol. Univ. Sao Paulo*, 1987, **1**(4), 9.
39. T. H. Grenby, *Pure Appl. Chem.*, 1997, **69**, 709.
40. N. L. M. Zanela, M. F. T. Borro Bijella and O. P. Da Silva Rosa, *Pesqui Odontol. Bras.*, 2002, **16**(2), 101.
41. S. L. Fischman, *J. Clin. Periodontol.*, 1986, **13**(5), 371, 379.
42. N. Chavarria, F. O. Gamboa, S. Escribano and R. Vitery, *Caries Res.*, 2008, **42**, 185.

43. S. B. M. Blauth de Slavutzky, *J. Verbraucherschutz Lebensmittelsicherh.*, 2010, **5**(2), 213.
44. V. A. Rumiantsev, V. V. Beliaev, V. A. Zubtsov, L. K. Esaian and I. V. Namestnikova, *Stomatologiia (Mosk)*, 2011, **90**(1), 18.
45. F. Gamboa and M. Chaves, *Acta Odontol. Latinoam.*, 2012, **25**(2), 171.
46. R. A. Giacaman, P. Campo, C. Muñoz-Sandoval and R. J. Castro, *Arch. Oral Biol.*, 2013, **58**(9), 1116.
47. F. C. Giongo, M. Bruna, C. C. Parolo, A. Carlén and M. Maltz, *Braz. Oral Res.*, 2014, **28**(1), 1.
48. S. L. Ajagannanavar, S. Shamarao, H. Battur, S. Tikare, A. A. Al-Kheraif and M. S. Al Sayed, *J. Int. Soc. Prev. Community Dent.*, 2014, 116.
49. M. S. Contreras, *J. Oral. Res.*, 2013, **2**(3), 158.
50. G. F. Ferrazzano, T. Cantile, B. Alcidi, M. Coda, A. Ingenito, A. Zarrelli and A. Pollio, *Molecules*, 2016, **21**(1), 38.
51. S. H. Lee, *J. Microbiol.*, 2013, **51**(4), 484.
52. K. Gawehn, in *Methods of Enzymatic Analysis*, ed. H. U. Bergmeyer, 3rd edn, 1984, vol. VI, p. 588.
53. S. Pierog, T. Hoppe, B. Schiermeyer, M. Frentzen and D. Kraus, *Proceedings of the 8th Stevia Symposium*, ed. J.M.C Geuns, Euprint, Heverlee, Belgium, 2015, p. 123.
54. A. Oehme, M. Wüst and U. Wölwer-Rieck, *Lebensmittelchemie*, 2013, **68**, 97.
55. K. Yasukawa, S. Kitanaka and S. Seo, *Biol. Pharm. Bull.*, 2002, **25**(11), 1488.
56. T. Akihisa, Y. Hamasaki, H. Tokuda, M. Ukiya, Y. Kimura and H. Nishino, *J. Nat. Prod.*, 2004, **67**(3), 407.
57. Y. Mizushina, T. Akihisa, M. Ukiya, Y. Hamasaki, C. Murakami-Nakai, I. Kuriyama, T. Takeuchi, F. Sugawara and H. Yoshida, *Life Sci.*, 2005, 77(17), 2127.
58. S. Jayaraman, M. S. Manoharan and S. Illanchezian, *Trop. J. Pharm. Res.*, 2008, 7(4), 1143.
59. M. Takasaki, T. Konoshima, M. Kozuka, H. Tokuda, J. Takayasu, H. Nishino, M. Miyakoshi, K. Mizutani and K. H. Lee, *Bioorg Med. Chem.*, 2009, **17**(2), 600.
60. P. Rajesh, V. R. Kannan and M. T. Durai, *J. Curr. Biotica.*, 2010, **3**(4), 549.

CHAPTER 10

Sensory Effects of Steviol Glycosides: Taste Perception and Beyond

KOENRAAD PHILIPPAERT[a,b,c], CAROLINE WUYTS[a,b], CAROLINE SIMOENS[a,b,d] AND RUDI VENNEKENS*[a,b]

[a]KU Leuven, Laboratory of Ion Channel Research, Department of Cellular and Molecular Medicine, Herestraat 49/802, 3000 Leuven, Belgium; [b]VIB Center for Brain & Disease Research, Leuven, Belgium; [c]Alberta Diabetes Institute, University of Alberta, 112 St NW, T6G 2T9, Edmonton, Canada; [d]KU Leuven, Laboratory of Clinical and Experimental Endocrinology, Department of Chronic Diseases, Metabolism and Ageing, Leuven, Belgium
*E-mail: Rudi.Vennekens@KULeuven.vib.be

10.1 Introduction

Steviol glycosides are found in the leaves of a scrub plant native to the sub-tropical regions of South America, *Stevia rebaudiana* Bertoni sp. Reports show that the biosynthesis and accumulation of steviol glycosides also occurs in two other plants: the Chinese blackberry *Rubus suavissimus* S.K. Lee, also known as Chinese sweet leaf and the Japanese *Angelica Keiskei* Koidz, or aka. Tomorrow's Leaf.[1–5] *S. rebaudiana* produces several steviol glycosides, of which stevioside and rebaudioside A are the most abundant.[6] They are based on a common diterpene aglycone core structure, steviol and are (poly)

Food Chemistry, Function and Analysis No. 7
Steviol Glycosides: Cultivation, Processing, Analysis and Applications in Food
Edited by Ursula Wölwer-Rieck

Published by the Royal Society of Chemistry, www.rsc.org

glycosylated on C_{13} or C_{19} with a variety of glucose moieties. The physiological pathway involved in the signal transduction of these molecules has only recently been identified. In this chapter, we aim to give an overview how steviol glycosides evoke their physiological effects.

10.2 Taste Aspects of Steviol Glycosides and Their Interaction with the Sweet-taste Receptor

Steviol glycosides evoke a sweet-taste sensation, which is the basis of the main economical use of steviol glycosides; as non-caloric sweeteners.[7] The drawback in the use of steviol glycosides as sweeteners is their lingering bitter aftertaste. It is a common observation that the sweet profile of high-intensity, non-caloric sweeteners is different compared to that of simple sugars as glucose and fructose. To understand why the taste profile of steviol glycosides differs from that of glucose, we need to understand the molecular interactions that occur during the perception of the taste of stevioside. It is important to make the differentiation between taste and hedonic perception of a compound. The former includes signalling that occurs through taste receptors in the taste buds; while the latter encompasses all sensory perception related to consumption of a product including taste, smell through the odorant receptors, signalling trough free trigeminal nerve endings in the oral cavity, post-ingestive effects of calorie consumption and, of course, personal appreciation.

10.2.1 Taste Perception

A primary taste is a sensation defined by the combination of six criteria: (1) it has an ecological consequence; (2) it is elicited by a distinctive class of chemicals; (3) specialised receptors exist and are activated; (4) primary taste is detected through gustatory nerves and processed in taste centres; (5) it is non-overlapping with other primary taste qualities; and (6) there exists a behavioural or physiological response.[8,9] There are currently five established taste modalities: salty, sour, bitter, sweet and umami; and recently more evidence is accumulating indicating fat taste as a sixth primary taste to detect fatty acids.[10–12] Ongoing research in the taste perception of complex carbohydrates aims to reveal if it can be considered a primary taste. It was observed that starch taste is, at least partially, independent of the sweet taste receptor in mice.[13] However, there are observed differences in taste perception between rodents and humans. Taste transduction plays a role in the evaluation of nutritious value, toxicity, sodium content and acidity of food.[14]

10.2.2 Taste Receptors

Most mammals express a range of taste receptors on specialised receptor cells organised in the taste buds on the tongue. Taste buds are structures containing 50–100 cells, subdivided into type I, type II and type III cells depending on their structure and gene expression. Type I cells are a heterogeneous group, have a glia-like function supporting the other taste bud cells. Type II cells are the receptor cells, they express one class of GPCR taste receptors from the Taste Receptor Type 1 (T1R) or Taste Receptor Type 2 (T2R) family, which responds to their sensitivity for either sweet/umami or bitter.[15] Type III cells are having the lowest prevalence in the taste buds and are responsible for the detection of sour taste.[16]

Sweet, bitter and umami taste perception is taste-receptor mediated. Type II taste cells express either one or more T2R receptors and respond to a variety of bitter substances, or they express T1R receptors and respond to sweet and/or umami substances. The T2R family consist of many members with an extensive genetic variation within species and in the number of T2R family genes between species, with 25–35 different T2R receptors in humans.[17,18] The T1R family is more conserved with generally three members that heterodimerise to form either the sweet (T1R2/T1R3) receptor or the umami (T1R1/T1R3) receptor. The relative specificity of the receptor cells is determined by the expression of the primary taste receptor. Type II cells share a common intracellular signal transduction pathway; the GPCR taste receptors activate heterotrimeric GTP-binding proteins (α-gustducin), resulting in functional interaction with $PLC\beta_2$, formation of IP_3, and release of Ca^{2+} from intracellular stores (Figure 10.1). Subsequently, the calcium activated monovalent cation channel Transient Receptor Potential channel subfamily Melastatin member 5 (TRPM5) is activated.[19,20] The activation of TRPM5 allows Na^+ influx and depolarisation of the membrane, a feature necessary for the subsequent release of the neurotransmitter ATP by the voltage sensitive CALHM1 channel.[21,22]

10.2.3 Taste Perception of Stevioside

Steviol glycosides have recently been approved for human consumption as a dietary supplement in the EU[23] and in the USA,[24] and are now commonly used in consumer beverages and other sweets, as replacement/reduction strategy for sugar. Stevioside and rebaudioside A taste approximately 300 times sweeter than sucrose.[7] However, the taste profile of steviol glycosides is somewhat different from sucrose. There is a slower onset and longer duration of the sweetness and there is a bitter aftertaste.[7,25]

It is currently not fully clear why steviol glycosides, such as stevioside and rebaudioside A, are so extremely sweet. Because of the inherent sweet taste and bitter aftertaste of steviol glycosides, many studies hypothesise that steviol glycosides interact with a combination of T2R bitter receptors and the T1R2 sweet receptor. Hellfritsch *et al.* showed that high concentrations of stevioside indeed activate human sweet-receptors, as well as certain bitter receptor in an overexpression system.[7] However, it was recently shown that stevioside and steviol modulate taste perception in T1R2/T1R3 knockout mice depending on the activity of TRPM5, which is an essential constituent in the downstream signalling of the taste receptor, as discussed above.[26]

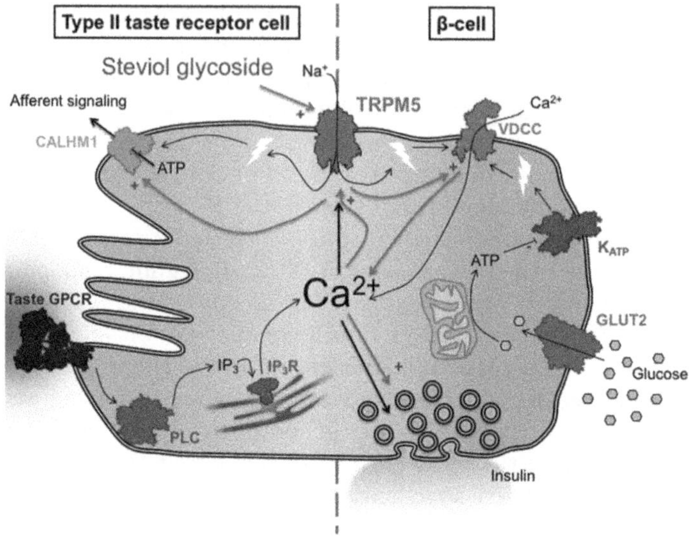

Figure 10.1 The role of TRPM5 in signal transduction of the type II taste receptor cell and insulin secreting β-cells. Important players in the signal transduction cascade leading to taste perception and insulin secretion are combined here in one scheme. The effects of stevioside through TRPM5 are highlighted with lighter coloured arrows. In type II taste receptor cells (left part of the scheme), activation of taste-specific (sweet, bitter, umami) G-protein coupled receptors leads to IP_3 mediated Ca^{2+} release and activation of TRPM5. The ensuing depolarisation activates CALHM1 channels that release ATP, which is the neurotransmitter that activates afferent signalling. Increased activity of TRPM5 likely increases ATP release and afferent signalling from the receptor cell, and thereby influences taste perception. In the pancreatic β-cell (right part of the scheme), glucose application induces closures of K_{ATP} channels, which leads to parallel V_M and $[Ca^{2+}]_{cyt}$ oscillations, which drive insulin secretion. TRPM5 is active during the lag phase in between bursts of action potentials and determines the frequency of V_M and $[Ca^{2+}]_{cyt}$ oscillations, which modulates insulin secretion: enhanced TRPM5 activity results in a higher oscillation frequency, which results in more insulin secretion. Note that steviol and its derivatives (*e.g.* stevioside) only enhance TRPM5 activity and insulin secretion, but that glucose is the trigger. Instead, insulinotropic compounds like glibenclamide (not depicted) that block K_{ATP} channels directly and trigger Ca^{2+} signalling and insulin secretion independent of glucose transport.

The interaction between the sweet taste receptor (T1R2, T1R3) and stevioside was shown through calcium imaging in an overexpression system.[7,27] On the T1R2 receptor, mutagenesis of D142 or E302 to alanine leads to a decreased stevioside-induced calcium influx in cell culture.[27] These residues are also essential for effects of sucrose on the sweet receptor.[28] *In silico* docking experiments on a predicted structure of T1R2, indicated the aforementioned residues to be interacting with stevioside. Docking of stevioside and rebaudioside A into a homology model of T1R3 based on the glutamate receptor reveals that stevioside and rebaudioside A have distinct binding orientations, suggesting that different amino acids are important for the interaction of either molecule with the sweet taste receptor.[29]

The sweet properties of steviol glycosides have been linked to a chemical moiety of the steviol core structure. Modifying the doubly bonded methylene on C_{17} to a ketone completely abolishes the sweet taste properties of both stevioside and rebaudioside A. Catalytic hydrogenation of rebaudioside B, rebaudioside C and rebaudioside D on C_{17} results in compounds with reduced but not absent sweet properties. This indicates that the C_{16}–C_{17} methylene double bond in steviol glycosides is an essential pharmacophore for the sweetness property of these molecules.[30,31] Based on these results, the main determinant for the sweetness of steviol glycosides would be the steviol moiety and not the glucose moieties. However, *in vivo* analysis of taste thresholds and post-threshold taste intensities clearly demonstrated the crucial influence of the number of β-glycosyl residues on the sweet taste elicited by individual steviol glycosides. Both glycone length at the separate glycosylation sites, and the total number of β-glycosyl moieties in the molecule are related to lower sweet thresholds and higher maximum sweet intensity. For example, rubusoside, stevioside, and rebaudioside A, which exhibit one β-glycosyl residue at C_{19}, show lower sweet threshold values and higher sweetness intensity as a function of the number of β-glycosyl residues at C_{13}.[7] Strikingly, in the same study it was shown that the efficacy of steviol glycosides in the cell-based assay was rather constant, whereas the maximal relative sweet intensities *in vivo* varied substantially. Hence, the distinct sweet intensities elicited by the various steviol glycosides seem not to be mediated at the level of the sweet taste receptor.[7]

Steviol glycosides also have a distinct bitter taste. The bitter taste of stevioside and rebaudioside A is mediated by two bitter receptors, *h*T2R4 and *h*T2R14. The bitter taste of steviol glycosides seems to be promoted by structural features that impair sweet taste. Marked differences in the total number of β-glycosyl moieties cause differences in bitter taste characteristics. Steviol glycosides bearing few β-glycosyl residues, such as rubusoside, showed lower bitter threshold values *in vivo* and higher bitter intensities compared to glycosides with many β-glycosyl residues, such as rebaudioside D.[7]

Recent work from our research group has shown that steviol glycosides potentiate activity of the TRPM5 ion channel, which is functionally coupled to both sweet and bitter taste receptors in taste receptor cells, but also other cell types, such as insulin secreting β-cells from the pancreas, or tuft cells in the gut. These findings profoundly change the current view how steviol glycosides influence taste.[26]

10.3 Steviol and Steviol Glycosides Act as a Taste Amplifier through Interaction with TRPM5

10.3.1 TRPM5

TRPM5 is a member of the TRPM subfamily of TRP channels.[32] It is a monovalent-cation-permeable ion channel in the plasma membrane and is activated by a rise of intracellular calcium.[33] It induces cellular depolarisation upon increases of $[Ca^{2+}]_i$. The channel is important in Ca^{2+} signalling and membrane excitability. Activation of TRPM5 is temperature-dependent as higher temperatures promote the activation of the channel, which is consistent with the observed temperature-dependent taste sensation mediated by TRPM5.[34,35]

Expression of TRPM5 was shown in several tissues throughout the body.[36,37] A genetic TRPM5 reporter mouse confirmed the expression of TRPM5 in distinct cell populations in several epithelia. TRPM5 positive cells were found in taste buds and in the olfactory epithelium.[38] A general observation is that TRPM5 is co-expressed with one or more signalling components of the canonical bitter, sweet and umami taste transduction cascade (PLCβ_2, α-gustducin). TRPM5 immunoreactivity is present throughout the body in cells suspected to play a role in chemosensation. TRPM5 expression is enriched in the vomeronasal organ and the olfactory epithelium, which is consistent with a proposed role for TRPM5 in the olfaction of odorants and pheromones, and general sensory perception.[39,40]

TRPM5 has a functional role in pancreatic β-cells where it is involved in the regulation of insulin secretion (Figure 10.1).[26,41,42] Additional TRPM5 expression has been shown in solitary chemosensory enteroendocrine cells, and brush cells in the respiratory system where the functional significance of TRPM5 needs to be further investigated.[37,39,43]

10.3.2 TRPM5 in Taste Sensation

Type II taste receptor cells in the taste buds express TRPM5 where it is acting as an essential part of the signal transduction cascade of bitter, sweet and umami taste receptors.[44] First indications of the physiological role of TRPM5 channels arose from the analysis of *Trpm5*$^{-/-}$ mice. These mice display a characteristic loss in the perception of sweet, bitter and umami taste. Deletion of TRPM5 results in an impaired or abolished sensitivity to bitter, sweet and umami tastes, although there is some debate whether some additional pathways, next to the TRPM5-dependent pathway, exist.[44–47] Indeed, whereas one study shows complete abolishment of behavioural responses to sweet, bitter and umami tastes in *Trpm5*$^{-/-}$ mice, other studies detect some residual activity in independently generated *Trpm5* knock-out mouse models.[44–46] Interestingly, *Trpm5*$^{-/-}$ mice have been shown to develop preferences for glucose over amino acid solutions independent of taste quality and caloric load, an effect associated with the ability of a given nutrient to regulate glucose metabolism and stimulate brain dopamine centres.[48]

As TRPM5 is an essential component of taste signalling, modulation of TRPM5 influences taste perception. TRPM5 channel activity can be modulated by heat: the channel is activated by increasing temperatures between 15 and 35 °C.[34] In line with this, TRPM5-dependent sweet taste responses in mouse gustatory nerves are strongly enhanced by heating, a feature lost in *Trpm5*$^{-/-}$ mice.[34] Temperature sensitivity of TRPM5 might underlie certain effects of taste, such as enhanced sweetness perception at high temperatures and thermal taste (a phenomenon whereby heating or cooling of the tongue evokes sensations of taste in the absence of tastants). Furthermore, the bitter tastant quinine dose-dependently inhibits TRPM5 currents and abolishes the gustatory response of sweet-sensitive gustatory nerves in WT but not in *Trpm5*$^{-/-}$ mice.[49] This inhibition of TRPM5 by bitter compounds constitutes the molecular basis of bitter-sweet taste interactions, whereby bitter tastants directly inhibit the sweet transduction pathway.

10.3.3 TRPM5 and Steviol Glycosides

Since the discovery of TRPM5 and its role in taste perception and insulin release, efforts have been made to modulate its channel activity with chemical compounds.[50]

Recently, steviol, stevioside and rebaudioside A were identified as potentiators of TRPM5.[26] It is currently unknown how TRPM5 and steviol glycosides interact. Neither the exact chemical determinant of the steviol glycosides nor the interacting amino acids of TRPM5 are known.

In taste preference tests, mice prefer a sweet tasting solution to water. This was shown in two bottle preference tests with sucrose, and the same is true for a solution containing stevioside or rebaudioside A. In the same test, a solution containing an equimolar concentration of steviol did not elicit a preference compared to water. Interestingly enough, the neutral-tasting steviol did enhance the preferences of sucrose, and increased the avoidance of a bitter compound.[26] This indicates that steviol acts as taste enhancer. This is in line with observations of a better sweet-taste profile of steviol glycosides when they are co-administered with glucose or sucrose, compared to the taste profile of steviol glycosides alone.[51]

When we used sweet receptor (T1R2/T1R3) knockout mice, there was no longer a taste preference for stevioside or rebaudioside A either in a 48 h two-bottle drinking assay, or a short access lickometer setup. These mice still show normal avoidance responses to the bitter tasting quinine. As in WT mice, the co-application of quinine and steviol increases avoidance of the bitter taste. Interestingly, the addition of stevioside to the quinine in sweet receptor deficient mice shows increased bitter avoidance, which strongly indicates a sweet-receptor independent action of stevioside on taste perception.[26] All the above described effects are absent in *Trpm5*$^{-/-}$ mice.

10.4 Anti-diabetic Effects of Steviol Glycosides through Their Interaction with TRPM5

10.4.1 Diabetes

Elevated blood glucose concentration is the hallmark symptom of diabetes, the most common metabolic disease in humans. The cause of diabetes is complex but generally stems from the inability of the pancreatic β-cells to provide sufficient insulin to maintain glycaemic control. This is either due to a partially deficient insulin secretion, an increased insulin resistance, or a combination of both. Type-1 diabetes is characterised by immune-mediated β-cell destruction and subsequent complete insulin deficiency.[52,53] Type-2 diabetes is a more complex disorder, with many genetic and environmental risk factors. There is a clear connection between age, a high-calorie diet and type-2 diabetes.[54] Lab animal studies revealed that a high-fat diet leads to gradual development of obesity,[55] glucose intolerance through insulin resistance and ultimately diabetes.[56]

Patients with type-2 diabetes are usually prescribed a combination of drugs and lifestyle modifications.[57] Most pharmaceuticals act on the insulin action, they either stimulate the secretion of insulin, reduce the insulin resistance or are mimicking insulin itself. Many of these drugs, like the sulfonylureas, come with associated risks on hypoglycaemic events and weight gain. Other drugs like incretin mimetics cause nausea and headaches, which leads to a lot of patients not adhering to their therapy.[58] There is a clear need for improvement of antidiabetic therapy, especially regarding prevention and progression of the disease.

10.4.2 TRPM5 and Diabetes

Apart from taste perception, TRPM5 also plays a role in the response of insulin-secreting pancreatic β-cells to sensing glucose in the blood. TRPM5 is expressed in insulin secreting β-cells.[26,42,59,60] The secretion of insulin by the pancreatic β-cell is a complex process driven by electrical activity and oscillations of the intracellular Ca^{2+} concentration.[61] Indeed, ATP produced by the glucose metabolism closes ATP-sensitive K^+ channels, allowing depolarisation of the membrane potential and activation of L-type Ca^{2+} channels. This leads to Ca^{2+} increase and secretion of insulin-containing vesicles.[62–65] As insulin is the only glucose-lowering hormone, its secretion is essential for glucose homeostasis.

A Ca^{2+}-activated non-selective monovalent cation channel could be measured in β-cells and was largely reduced in *Trpm5*$^{-/-}$ mice.[42,66] TRPM5 functions as a positive regulator of glucose-induced insulin release. This is reflected by observations in *Trpm5*$^{-/-}$ mice, where glucose-induced insulin release is impaired but not completely absent. Potentiation of TRPM5 activity on the other hand increases insulin secretion.[26] TRPM5-deficient mice display an impaired glucose tolerance during oral and intraperitoneal glucose tolerance tests.[26,41,42] This prediabetic phenotype is caused by a partial β-cell dysfunction and leads to an increased development of diabetes during conditions of higher insulin demand (such as pregnancy, obesity, ageing, *etc.*).[26]

Colsoul *et al.* showed that TRPM5 determines the frequency of glucose-induced Ca^{2+} oscillations of pancreatic islets (Figure 10.1).[42] In agreement with this, the application of stevioside resulted in a dose-dependent increase of the frequency of glucose-induced Ca^{2+} oscillations in WT islets. Similar results were obtained after the application of steviol. Neither steviol nor stevioside affected the oscillatory behaviour of *Trpm5*$^{-/-}$ islets. These findings demonstrate that steviol glycosides increase the frequency of glucose-induced calcium oscillations in pancreatic islets in a TRPM5-dependent manner.[26] Considering that Ca^{2+} oscillation frequency correlates with insulin release, stevioside potentiated glucose-induced insulin release in WT islets, but not in *Trpm5*$^{-/-}$ islets.[26]

Not much is known about a possible role for TRPM5 in the secretion of other hormones by the pancreatic islet. Currently, it cannot be excluded that a paracrine effect within the islets modulates the perceived function of TRPM5 in the β-cells. Next to the insulin-secreting β-cells, the islets of Langerhans comprise several other cell types such as glucagon-releasing α-cells and somatostatin-producing δ-cells. TRPM5 is shown to co-localise with insulin in mouse β-cells with immunostaining. Q-PCR on purified α- or β-cells shows significantly more TRPM5 in the β-cell fraction.[42] Furthermore in differentiation studies, it was found that TRPM5 is the third most upregulated gene in β-cells compared to the expression in α-cells, indicating TRPM5 is a legitimate marker of β-cells in the pancreatic islet.[59] This makes it reasonable that any of the TRPM5-mediated effects observed in islets are due to the function of TRPM5 in the insulin secreting β-cells.

10.4.3 Steviol Glycosides and Diabetes

Several health benefits have been ascribed to steviol glycosides. Stevioside, and/or stevia-extract, have anti-hypertensive, anti-inflammatory, anti-cancer effects, and anti-diarrheal effects. However, a complication in many of these studies is often the purity and composition of the stevia extract that is used. This results in a conflict and a difficult-to-interpret set of data. Nevertheless, a large effort went into the investigation of the effects of stevia extract, steviol glycosides, steviol and derivatives on insulin secretion and carbohydrate metabolism. A systematic analysis of anti-diabetic effects of steviol glycosides was performed by Jeppesen, Hermansen and colleagues. They found that stevioside and steviol dose-dependently enhanced insulin release from mouse islets and the INS-1 cell line in the presence of 16.7 mM glucose. This effect was critically dependent on the prevailing glucose concentration, as both compounds only potentiated insulin secretion at or above 8.3 mM.[67] Later similar effects were observed with rebaudioside A on murine pancreatic islets.[68]

Our research group showed recently that in living mice, blood glucose values were significantly reduced after stevioside treatment in WT but not *Trpm5*$^{-/-}$ mice. This effect was even more pronounced in obese WT mice, after 20 weeks of high fat diet (HFD). Notably, both in WT and *Trpm5*$^{-/-}$ animals, we observed no effect of stevioside administration on fasting glucose level, indicating that stevioside by itself does not induce hypoglycaemia. Even more striking, daily stevioside supplementation prevented the development of chronic hyperglycaemia in mice fed with a HFD.[26] Animals treated with stevioside maintained a normal glycaemic profile over the course of 20 weeks of HFD, while non-treated mice on the same diet developed hyperglycaemia, as expected. In *Trpm5*$^{-/-}$ animals, no difference between the control and the stevioside treated group was observed. Notably, weight gain and the development of insulin resistance was similar in stevioside-treated and non-treated mice. Furthermore, withdrawal of mice from stevioside treatment quickly reversed the beneficial effect on hyperglycaemia. Taken together, these data indicate that stevioside promotes glucose-induced insulin secretion through its interaction with TRPM5, and represents in this way a completely novel type of insulinotropic medication, which might be helpful for the prevention and treatment of type-2 diabetes.[26]

Several studies have reported on the *in vivo* effects of steviol and steviol glycosides in animal models and humans. In Goto-Kakizaki rats (a non-obese Wistar rat substrain, which develops type-2 diabetes early in life), acute and chronic stevioside treatment decreased the increase of plasma glucose, increased the insulin response and suppressed the glucagon level in a *i.v.* glucose tolerance test.[69,70] Furthermore, in this set-up, treated animals displayed a significantly decreased systolic and diastolic blood pressure (±10% lower compared to control). In contrast, the same dose of rebaudioside A did not have any effects on glycaemia in the same animal model. In Zucker diabetic fatty rats (an obese, hereditary type-2 diabetic animal model), stevioside decreased glycemia in an arterial glucose tolerance test, decreased fasting glucose levels and decreased blood pressure.[71,72] In *Ob*/*Ob* and Low Density Lipoprotein Receptor double deficient mice, stevia-derived compounds (stevioside, rebaudioside A and steviol) reduce hepatic steatosis and improved glucose metabolism, fat catabolism, bile acid metabolism, and lipid storage and transport.[73] Steviol glycosides have also been implicated to modulate glucose transport in several cell types.[74] Isosteviol, derived from steviol by acidic hydrolysis, was shown to have antidiabetic properties. It reduces plasma glucose levels in an *i.v.* glucose tolerance test in Zucker diabetic fatty rats.[75] In KKAy mice (a mouse line characterised by obesity, hyperglycaemia, hyperinsulinemia and glucose intolerance) isosteviol increased glucose and insulin sensitivity as well as the lipid profile and upregulates the gene expression of key β-cell genes, including insulin regulatory transcription factors.[76]

10.5 Anti-hyperglycaemic Effects of Steviol Glycosides in Humans

The overall goal of diabetes management is focused on maintaining blood glucose levels as close to normal as possible, without risking hypoglycaemic events, in order to prevent or delay hyperglycaemia-associated complications. Unlike in animals, a clear conclusion on the effect of steviol glycosides on glucose homeostasis in humans is lacking. A limited number of clinical trials have been conducted in healthy volunteers and individuals with type-2 diabetes mellitus (T2DM) to evaluate the effect of stevia extracts on glucose homeostasis following single or repeated administration. A summary of conducted trails with stevia-related compounds is given in Table 10.1.

10.5.1 Studies with Healthy Volunteers

An early study in humans investigated the effect of aqueous extracts of *Stevia rebaudiana* leaves on a glucose tolerance test (GTT). Administration of 5 g extracts of leaves to healthy volunteers at 6 h intervals for 3 d resulted in significantly lower plasma glucose levels than the control condition, at each time point tested. The maximum difference in plasma glucose levels for baseline and following stevia administration was observed at 30 min after the glucose challenge, *i.e.* the highest glucose level during the GTT.[77] In contrast to this outcome, two other performed studies in healthy volunteers were unable to confirm the glucose-lowering effect of stevia extracts. Geuns *et al.* gave 250 mg stevioside three times daily for 3 d orally and collected blood samples after fasting, before and at different time points at the third day of stevioside administration. Mean blood glucose and insulin were not significantly different between the control and stevioside condition.[78] In the double-blind, placebo-controlled long-term study of Barriocanal *et al.*, subjects were randomly allocated to active treatment or placebo and followed-up for 3 months. The repeated use of 250 mg steviol glycosides per day during 3 months did not change glucose or glycated haemoglobin (HbA1c) levels significantly from baseline measurements.[79]

10.5.2 Studies with Individuals Diagnosed with Type-2 Diabetes Mellitus

Enhanced insulin secretion upon impaired response to glucose stimulation is the intended mechanism of action for steviol glycosides. This makes subjects with type-2 diabetes mellitus very suitable to provide data on the effects of steviol glycosides on glucose homeostasis. A limited

Table 10.1 An overview of clinical trials using steviol glycosides on healthy subjects or patients with type-2 diabetes. A brief summary of the main findings and the methodology of the studies is indicated.

First author, year	Design	Participants	Compound	Dose	Administration	Duration	Outcome	Results
Barriocanal, 2008	Randomised, double-blind, parallel	Healthy volunteers ($N = 30$)	Steviol glycoside (purity ≥ 92%)	250 mg 3 times per day	Oral (capsule)	3 months	Δ from baseline	Glucose *steviol glycoside* 82.5 to 82.9 mg dL^{-1}; Glucose *placebo* 82.9 to 83.9 mg dL^{-1} Insulin *steviol glycoside* 4.2 to 4.9 μUI mL^{-1}; Insulin *placebo* 8.4 to 8.7 μUI mL^{-1} HbA1c *steviol glycoside* 5.3–5.6%; HbA1c *placebo* 5.3–5.4%
		Individuals with T2DM ($N = 30$)					Δ from baseline	Glucose *steviol glycoside* 151.2 to 133.8 mg dL^{-1}; Glucose *placebo* 131.3 to 118.9 mg dL^{-1} Insulin *steviol glycoside* 13.3 to 11.6 μUI mL^{-1}; Insulin *placebo* 14.7 to 15.3 μUI mL^{-1} HbA1c *steviol glycoside* 6.8–6.6%; HbA1c *placebo* 6.8–6.8%
Curi, 1986	Open label	Healthy volunteers ($N = 16$)	Extract of stevia rebaudiana leaves	5 g extract 13 doses/3 days	Oral	3 days	Plasma glucose	Lower after Stevia treatment compared to control, for each time tested

Geuns, 2007	Open label	Healthy volunteers ($N = 9$)	Stevioside (97%), steviolbioside (2.7%), rebaudioside A (0.3%)	250 mg 3 times per day	Oral (capsule)	3 days	Fasting glucose	Stevioside: 4.72 ± 0.08 mM; control: 4.63 ± 0.12 mM
							Glucose *30 min*	Stevioside: 4.63 ± 0.11 mM; control: 4.65 ± 0.09 mM
							Glucose *60 min*	Stevioside: 4.55 ± 0.09 mM; control: 4.51 ± 0.11 mM
							Fasting insulin	Stevioside: 6.43 ± 0.85 mM; control: 6.79 ± 0.78 mM
							Insulin *30 min*	Stevioside: 5.67 ± 0.81 mM; control: 4.93 ± 0.96 mM
							Insulin *60 min*	Stevioside: 5.62 ± 0.65 mM; control: 5.08 ± 0.36 mM
Gregersen, 2004	Randomised, paired, cross-over	Individuals with T2DM ($N = 12$)	Extracts of stevia rebaudiana leaves (91% stevioside, 4% rebaudioside A, 5% other stevioside derivates)	1 g stevioside	Oral (capsule) With test meal	N.A.	Postprandial blood glucose	Reduced by 18% ± 5% in stevioside ($P < 0.004$) compared to placebo (AUCi, glucose, control 638 ± 55 *vs.* AUCi, glucose, stevioside 522 ± 64 mmol $L^{-1} \times 240$ min)
							Postprandial insulin	AUCi, insulin, control 34 991 ± 3550 *vs.* 38 499 ± 3740 pmol L^{-1} ($P = 0.08$)
							Insulinogenic index	Control 60 ± 8 *vs.* stevioside 84 ± 11 ($P < 0.001$)

(*continued*)

Table 10.1 (*continued*)

First author, year	Design	Participants	Compound	Dose	Administration	Duration	Outcome	Results
Maki, 2008	Randomised, double-blind, placebo-controlled	Individuals with T2DM ($N = 122$)	Rebaudioside A (97% purity)	250 mg 4 times per day	Oral (capsule)	16 weeks	HbA1c	No significant difference between treatment groups from baseline to 4, 8, 12 and 16 weeks
							Δ fasting glucose from baseline to mean value of 12–16 weeks	7.5 ± 3.7 mg dL^{-1} for rebaudioside A; 11.2 ± 4.5 mg dL^{-1} for placebo ($p = 0.966$)
							Δ fasting insulin from baseline to mean value of 12–16 weeks	1.0 ± 0.6 μU mL^{-1} for rebaudioside A; 3.3 ± 1.5 μU mL^{-1} for placebo ($p = 0.979$)
							Δ fasting C-peptide from baseline to mean value of 12–16 weeks	0.1 ± 0.1 ng mL^{-1} for rebaudioside A; 0.4 ± 0.1 ng mL^{-1} for placebo ($p = 0.313$)

number of studies have been performed to evaluate the effect of stevia extracts on blood glucose in type-2 diabetic patients. In a first study, stevioside suppressed the postprandial blood glucose on average by 18% in comparison to placebo and tended to enhance insulin secretion, when administered in a dose of 1 g with a meal.[80] Moreover, the insulinogenic index, indicating insulin secretion, increased significantly by approximately 40% after stevioside treatment compared to control. Postprandial GLP-1 and GIP levels did not differ significantly for both conditions while glucagon levels were significantly reduced in the stevioside condition. Despite the positive outcome on blood glucose levels, the trial included a low number of participants with subsequent insufficient statistical power and allowed oral anti-diabetic agents that might interfere with the clinical endpoints in the trial. More recent examinations have not provided clear evidence to support the previous results. In the study of Barriocanal *et al.*, there was no significant change in glucose or HbA1c after 250 mg stevioside treatment three times daily for 3 months compared to the baseline.[79] In another long-term study, subjects with T2DM were randomised to receive four capsules of 250 mg rebaudioside A or placebo for 16 weeks following a two-week single-blind placebo lead-in period. The mean change in HbA1c throughout the study was not significantly different for rebaudioside A administration *vs.* placebo.[81] The absence of any effect can be explained by the permitted use of hypoglycaemic agents or insulin therapy that influences the outcome.

Most studies available on steviol glycosides have been performed in either cell cultures, isolated tissues or experimental animals, while only few studies have been carried out in humans. Differences in testing protocols results moreover in an apparent dissimilarity on the effects in these clinical trials which makes it challenging to draw conclusions.[82] Ultimately, individuals with T2DM are acknowledged to be more susceptible to blood-glucose lowering effects of stevia extracts. In that context the Food Standards Australia and New Zealand Foods (FSANZ) proposes that *"The effects of steviol glycosides only occur when blood glucose concentrations are elevated, as in the diabetic state, and that there is a relatively low risk of hypoglycaemia in normal subjects from consumption of dietary concentrations of stevioside"*.[82] Clearly, more research in humans is needed to test the fundamental aspect of steviol glycosides on blood glucose homeostasis, certainly in individuals coping with T2DM.

10.6 Concluding Remarks

The medicinal use of natural products, extracted from plants, animals or micro-organisms, precedes the recorded human history. Scientists have been able to elucidate the working mechanisms of several of these natural compounds and isolate the active molecules to use as pharmaceuticals. In the last decades the effects of stevioside and related steviol glycosides on animal and human physiology have been extensively studied. No study shows detrimental effects on survival or fitness, using physiological concentrations of steviol glycosides.[83–85] Next to the investigation on the flavour profile of the glycosides, studies investigating the antidiabetic effects are arguably the most prevalent. Recently, TRPM5 was found to be essential for mediating the effects of stevioside, both on taste perception and insulin secretion. This elucidation of the molecular pathway through which stevioside interacts with the body indicates that stevioside can be a valid candidate to be used in the treatment of type-2 diabetes.

Acknowledgements

This work was supported by FWO Vlaanderen (G.0761.10N, G.0596.12 and G.0565.07), IUAP program from the Belgian Federal Government (Ice-Path, P7/13), KU Leuven 'Bijzonder Onderzoeksfonds' (STRT1/09/046, GOA 2009/07, EF/95/010 and TRPLe) and the EFSD (EFSD/Novo Nordisk Programme for Diabetes Research in Europe). KP is a FWO [PEGASUS]^2 Marie Skłodowska-Curie Fellow under the Marie Skłodowska-Curie grant agreement No 665501.

References

1. T. Tanaka, H. Kohda, O. Tanaka, F.-H. Chen, W.-H. Chou and J.-L. Leu, *Agric. Biol. Chem.*, 1981, **45**, 2165.
2. S. Ceunen, D. B. Wim, F. Compernolle, A. H. Mai and J. M. C. Geuns, *Carbohydr. Res.*, 2013, **379**, 1.
3. L. Jun-fang, C. Ya-hui, Z. Yan-chun, D. Hai-rong and H. Fei-long, *Chin. J. Exp. Tradit. Med. Formulae*, 2012, **18**, 103.
4. A. D. Kinghorn, D. D. Soejarto, N. P. Nanayakkara, C. M. Compadre, H. C. Makapugay and J. M. Hovanec-Brown, *et al.*, *J. Nat. Prod.*, 1984, **47**, 439.
5. S. Ceunen and J. M. C. Geuns, *J. Nat. Prod.*, 2013, **76**, 1201.
6. R. S. Pawar, A. J. Krynitsky and J. I. Rader, *Anal. Bioanal. Chem.*, 2013, **405**, 4397.
7. C. Hellfritsch, A. Brockhoff, F. Stähler, W. Meyerhof and T. Hofmann, *J. Agric. Food Chem.*, 2012, **60**, 6782.
8. K. Kurihara and M. Kashiwayanagi, *Ann. N. Y. Acad. Sci.*, 1998, **855**, 393.
9. R. D. Mattes, *Physiol. Behav.*, 2011, **104**, 624.
10. R. S. Keast and A. Costanzo, *Flavour*, 2015, **4**, 5.
11. C. A. Running, B. A. Craig and R. D. Mattes, Oleogustus, *Chem. Senses*, 2015, **40**, 507.
12. R. M. Tucker, K. A. Kaiser, M. A. Parman, B. J. George, D. B. Allison and R. D. Mattes, *PLoS One*, 2017, **12**, 1.
13. S. Zukerman, J. I. Glendinning, R. F. Margolskee and A. Sclafani, *Chem. Senses*, 2013, **38**, 421.
14. N. Chaudhari and S. D. Roper, *J. Cell Biol.*, 2010, **190**, 285.
15. S. D. Roper and N. Chaudhari, *Nat. Rev. Neurosci.*, 2017, **18**, 485.

16. A. L. Huang, X. Chen, M. A. Hoon, J. Chandrashekar, W. Guo and D. Tränkner, *et al.*, *Nature*, 2006, **442**, 934.
17. W. Meyerhof, S. Born, A. Brockhoff and M. Behrens, *Flavour Fragrance J.*, 2011, **26**, 260.
18. P. Shi and J. Zhang, *Mol. Biol. Evol.*, 2006, **23**, 292.
19. Z. Zhang, Z. Zhao, R. Margolskee and E. Liman, *J. Neurosci.*, 2007, **27**, 5777.
20. T. Hofmann, V. Chubanov, T. Gudermann and C. Montell, *Curr. Biol.*, 2003, **13**, 1153.
21. A. Taruno, V. Vingtdeux, M. Ohmoto, Z. Ma, G. Dvoryanchikov and A. Li, *et al.*, *Nature*, 2013, **495**, 223.
22. Z. Ma, a. P. Siebert, K.-H. Cheung, R. J. Lee, B. Johnson and a. S. Cohen, *et al.*, *Proc. Natl. Acad. Sci. U. S. A.*, 2012, **109**, E1963.
23. The European Commission, Amending Annex II to Regulation (EC) No 1333/2008 of the European Parliament and of the Council with regard to steviol glycosides, *Off. J. Eur. Union*, 2011, **L 295**, 205.
24. FDA. Agency Response Letter GRAS Notice No. GRN 000395. GRAS Not Invent 2012; GRN 00395. http://www.fda.gov/Food/IngredientsPackagingLabeling/GRAS/NoticeInventory/ucm291295.htm.
25. D. Risso, G. Morini, L. Pagani, A. Quagliariello, C. Giuliani and S. De Fanti, *et al.*, *Genes Nutr.*, 2014, **9**, 401.
26. K. Philippaert, A. Pironet, M. Mesuere, W. Sones, L. Vermeiren and S. Kerselaers, *et al.*, *Nat. Commun.*, 2017, **8**, 14733.
27. F. Zhang, B. Klebansky, R. M. Fine, H. Liu, H. Xu and G. Servant, *et al.*, *Proc. Natl. Acad. Sci. U. S. A.*, 2010, **107**, 4752.
28. S.-K. Kim, Y. Chen, R. Abrol, W. A. Goddard and B. Guthrie, *Proc. Natl. Acad. Sci. U. S. A.*, 2017, **114**, 2568.
29. J. Ahmed, S. Preissner, M. Dunkel, C. L. Worth, A. Eckert and R. Preissner, *Nucleic Acids Res.*, 2011, **39**, 1.
30. I. Prakash, M. Campbell and V. S. P. Chaturvedula, *Int. J. Mol. Sci.*, 2012, **13**, 15126.
31. M. Upreti, G. Dubois and I. Prakash, *Molecules*, 2012, **17**, 4186.
32. X. Z. Xu, F. Moebius, D. L. Gill and C. Montell, *Proc. Natl. Acad. Sci. U. S. A.*, 2001, **98**, 10692.
33. D. Prawitt, M. K. Monteilh-Zoller, L. Brixel, C. Spangenberg, B. Zabel and A. Fleig, *et al.*, *Proc. Natl. Acad. Sci. U. S. A.*, 2003, **100**, 15166.
34. K. Talavera, K. Yasumatsu, T. Voets, G. Droogmans, N. Shigemura and Y. Ninomiya, *et al.*, *Nature*, 2005, **438**, 1022.
35. K. Talavera, Y. Ninomiya, C. Winkel, T. Voets and B. Nilius, *Cell. Mol. Life Sci.*, 2007, **64**, 377.
36. T. Enklaar, M. Esswein, M. Oswald, K. Hilbert and H. M. Winterpacht a, *et al.*, *Genomics*, 2000, **67**, 179.
37. C. a Pérez, L. Huang, M. Rong, J. A. Kozak, A. K. Preuss and H. Zhang, *et al.*, *Nat. Neurosci.*, 2002, **5**, 1169.
38. S. Kusumakshi, A. Voigt, S. Hubner, I. Hermans-Borgmeyer, A. Ortalli and M. Pyrski, *et al.*, *Chem. Senses*, 2015, **40**, 413.

39. S. Kaske, G. Krasteva, P. König, W. Kummer, T. Hofmann and T. Gudermann, *et al.*, *BMC Neurosci.*, 2007, **8**, 49.
40. A. Oshimoto, Y. Wakabayashi, A. Garske, R. Lopez, S. Rolen and M. Flowers, *et al.*, *PLoS One*, 2013, **8**, e61990.
41. L. R. Brixel, M. K. Monteilh-Zoller, C. S. Ingenbrandt, A. Fleig, R. Penner and T. Enklaar, *et al.*, *Pflugers Arch.*, 2010, **460**, 69.
42. B. Colsoul, A. Schraenen, K. Lemaire, R. Quintens, L. Van Lommel and A. Segal, *et al.*, *Proc. Natl. Acad. Sci. U. S. A.*, 2010, **107**, 5208.
43. Z. Kokrashvili, B. Mosinger and R. F. Margolskee, *Am. J. Clin. Nutr.*, 2009, **90**, 822S.
44. Y. Zhang, M. a Hoon, J. Chandrashekar, K. L. Mueller, B. Cook and D. Wu, *et al.*, *Cell*, 2003, **112**, 293.
45. M. C. Eddy, B. K. Eschle, D. Peterson, N. Lauras, R. F. Margolskee and E. R. Delay, *Chem. Senses*, 2012, **37**, 391.
46. S. Damak, M. Rong, K. Yasumatsu, Z. Kokrashvili, C. A. Pérez and N. Shigemura, *et al.*, *Chem. Senses*, 2006, **31**, 253–264.
47. D. Dutta Banik, L. E. Martin, M. Freichel, A.-M. Torregrossa and K. F. Medler, *Proc. Natl. Acad. Sci. U. S. A.*, 2018, 201718802.
48. X. Ren, J. G. Ferreira, L. Zhou, S. J. Shammah-Lagnado, C. W. Yeckel and I. E. de Araujo, *J. Neurosci.*, 2010, **30**, 8012.
49. K. Talavera, K. Yasumatsu, R. Yoshida, R. F. Margolskee, T. Voets and Y. Ninomiya, *et al.*, *FASEB J.*, 2008, **22**, 1343.
50. K. Philippaert, S. Kerselaers, T. Voets and R. Vennekens, *SLAS Discov. Adv. Life Sci. R&D*, 2018, **23**, 341.
51. G. E. DuBois and I. Prakash, *Annu. Rev. Food Sci. Technol.*, 2012, **3**, 353.
52. L. Brunton and B. K. Bruce Chabner, *Goodman and Gilman's The Pharmacological Basis of Therapeutics. Twelfth Edition*, McGraw Hill Professional, 2011.
53. J. Thompson Coon, *Goodman and Gilman's the Pharmacological Basis of Therapeutics*, McGraw-Hill, 2010, DOI: 10.1111/j.2042-7166.2002.tb05480.x.
54. C. Herder and M. Roden, *Eur. J. Clin. Invest.*, 2011, **41**, 679.
55. N. Hariri and L. Thibault, *Nutr. Res. Rev.*, 2010, **23**, 270.
56. M. S. Winzell and B. Ahrén, *Diabetes*, 2004, **53**(3), S215.
57. H. E. Lebovitz, *Nat. Rev. Endocrinol.*, 2011, **7**, 408.
58. S. V. Joy, P. T. Rodgers and A. C. Scates, *Ann. Pharmacother.*, 2005, **39**, 110.
59. H. Chakravarthy, X. Gu, M. Enge, X. Dai, Y. Wang and N. Damond, *et al.*, *Cell Metab.*, 2017, **25**, 622.
60. R. Vennekens, M. Mesuere and K. Philippaert, *Acta Physiol.*, 2018, **222**, e12949.
61. P. E. MacDonald and P. Rorsman, *PLoS Biol.*, 2006, **4**, 167.
62. G. Drews, P. Krippeit-Drews and M. Düfer, *Adv. Exp. Med. Biol.*, 2010, **654**, 115.
63. J. C. Henquin, M. Nenquin, M. A. Ravier and A. Szollosi, *Diabetes, Obes. Metab.*, 2009, **11**(Suppl 4), 168.
64. J. C. Henquin, *Diabetologia*, 2009, **52**, 739.

65. P. E. MacDonald and P. Rorsman, *Physiology*, 2007, **22**, 113.
66. D. Liu and E. R. Liman, *Proc. Natl. Acad. Sci. U. S. A.*, 2003, **100**, 15160.
67. P. B. Jeppesen, S. Gregersen, C. R. Poulsen and K. Hermansen, *Metabolism*, 2000, **49**, 208.
68. R. Abudula, P. B. Jeppesen, S. E. D. Rolfsen, J. Xiao and K. Hermansen, *Metabolism*, 2004, **53**, 1378.
69. P. B. Jeppesen, S. E. Dyrskog, A. Agger, S. Gregersen, M. Colombo and J. Xiao, *et al.*, *Rev. Diabet. Stud.*, 2006, **3**, 189.
70. P. B. Jeppesen, S. Gregersen, S. E. D. Rolfsen, M. Jepsen, M. Colombo and a. Agger, *et al.*, *Metabolism*, 2003, **52**, 372.
71. S. E. U. Dyrskog, P. B. Jeppesen, M. Colombo, R. Abudula and K. Hermansen, *Metabolism*, 2005, **54**, 1181.
72. N. Lailerd, V. Saengsirisuwan, J. A. Sloniger, C. Toskulkao and E. J. Henriksen, *Metabolism*, 2004, **53**, 101.
73. P. Holvoet, A. Rull, A. García-Heredia, S. López-Sanromà, B. Geeraert and J. Joven, *et al.*, *Food Chem. Toxicol.*, 2015, **77**, 22.
74. B. Rizzo, L. Zambonin, C. Angeloni, E. Leoncini, F. Vieceli Dalla Sega and C. Prata, *et al.*, *Oxid. Med. Cell. Longevity*, 2013, **2013**, ID348169, DOI: 10.1155/2013/348169.
75. J. Ma, Z. Ma, J. Wang, R. W. Milne, D. Xu and a. K. Davey, *et al.*, *Diabetes, Obes. Metab.*, 2007, **9**, 597.
76. I. Nordentoft, P. B. Jeppesen, J. Hong, R. Abudula and K. Hermansen, *Diabetes, Obes. Metab.*, 2008, **10**, 939.
77. R. Curi, M. Alvarez, R. B. Bazotte, L. M. Botion, J. L. Godoy and A. Bracht, *Braz. J. Med. Biol. Res.*, 1986, **19**, 771.
78. J. M. C. Geuns, J. Buyse, A. Vankeirsbilck and E. H. M. Temme, *Exp. Biol. Med.*, 2007, **232**, 164.
79. L. a Barriocanal, M. Palacios, G. Benitez, S. Benitez, J. T. Jimenez and N. Jimenez, *et al.*, *Regul. Toxicol. Pharmacol.*, 2008, **51**, 37.
80. S. Gregersen, P. B. Jeppesen, J. J. Holst and K. Hermansen, *Metabolism*, 2004, **53**, 73.
81. K. C. Maki, L. L. Curry, M. S. Reeves, P. D. Toth, J. M. McKenney and M. V. Farmer, *et al.*, *Food Chem. Toxicol.*, 2008, **46**(Suppl 7), S47.
82. FSANZ. Final Assessment Report Application A540 Steviol Glycosides as Intense Sweeteners. 2008.
83. H. Mitsuhashi, *Tama. Biochem. Co. Ltd*, 1976, 1.
84. C. Toskulkao, L. Chaturat, P. Temcharoen, T. Glinsukon, C. Toskulkac and L. Chaturat, *et al.*, *Drug Chem. Toxicol.*, 1997, **20**, 31.
85. V. Yodyingyuad and S. Bunyawong, *Hum. Reprod.*, 1991, **6**, 158.

Subject Index